Tran

Questions and Answers books are available on the following subjects

QUESTIONS & ANSWERS

Transistors

Ian R. Sinclair

Newnes Technical Books

Newnes Technical Books

is an imprint of the Butterworth Group
which has principal offices in
London, Sydney, Toronto, Wellington, Durban and Boston

First published 1966 by George Newnes Ltd
Second edition 1967
Third edition 1969
   Reprinted 1971, 1976
Fourth edition 1980 by Newnes Technical Books

British Library Cataloguing in Publication Data

Sinclair, Ian
   Transistors. – 4th ed. – (Questions & answers).
   1. Transistors
   I. Title   II. Brown, Clement   III. Questions
   and answers on transistors
   621. 3815'28   TK7871.9   80-40378

   ISBN 0-408-00485-1

Typeset by Butterworths Litho Preparation Department
Printed in England by Butler & Tanner Ltd, Frome and London

PREFACE

In a few years 'transistor' has become a household word – the word being often used, understandably but incorrectly, to describe a portable radio using transistors. Today, the transistor has assumed a dominant position in electronics, being used in vast numbers in radio receivers and other domestic electronic equipment, computers, instruments, industrial control equipment, data processing equipment, air and sea navigational and communications equipment, telecommunications links and so on. Hence the wide interest in these small devices that have made so much possible. The aim of this short book is to explain the basic features of transistors, how they work and what they can do, and survey the many applications in which they are used.

The transistor is but one of the many devices that make use of the electrical characteristics of semiconductor material. Therefore in addition to the transistor opportunity has been taken to describe a number of related devices, many of which are commonly used in conjunction with transistors.

More recently there have been profound changes in the extent to which transistors have replaced valves and have themselves in turn been replaced by integrated circuits. These changes have necessitated very considerable rewriting for this edition, but I hope that it still remains true to Clement Brown's original book which has been so very useful in the past.

Much useful information has been provided by Brush Electrical Engineering Co. Ltd., Ever Ready Co. Ltd., Hird-Brown Ltd., I.B.M. United Kingdom Ltd., M.E.L. Equipment Co. Ltd., Mullard Ltd. and Newmarket Transistors Ltd. Their co-operation is gratefully acknowledged.

Fig. 55 is reproduced by kind permission from *Practical Electronics*.

Ian Sinclair

CONTENTS

1

SEMICONDUCTORS

What is a transistor?

It is a device which can switch and amplify electrical signals. The transistor has three terminals, called emitter, base and collector. A small current flowing between the base and the emitter can control a much larger current (some 25 to 1000 times larger) flowing between the collector and the emitter. The action

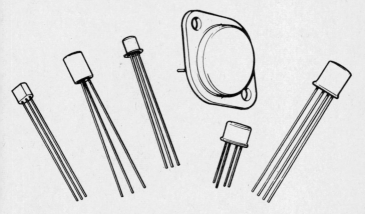

Fig. 1. Various sizes and shapes of transistors

of the transistor depends on the properties of semiconductor materials. Bipolar transistors make use of 'junctions' formed by growing one type of semiconductor material on top of another.

Field-effect transistors (fets) are constructed in a different way. A few typical transistor types are shown in Fig. 1.

What are semiconductors?

A large number of materials, intermediate between conductors and insulators, come under the heading of semiconductor. Most metals readily conduct electricity, and this is due to the presence in the of very large numbers of free electrons. These are electrically negative; move under the influence of voltage and act as carriers of current. On the other hand, in an insulator (for example mica, or one of the plastics used in electrical work) few electrons are free to move, and therefore no current – or very little – flows if a voltage is applied.

The availability of free electrons depends on the material's atomic structure, the relative positions of atoms, and the temperature. Especially important in the present context is the fact that a small number of 'foreign' atoms can influence the electrical properties of certain materials. These materials are the semiconductors, the most important of which, in present-day transistor manufacture, are germanium and silicon. In general, semiconductors conduct more readily with increase in temperature; the converse is true of good conductors, such as copper.

What is the origin of the electrons?

Electrons are part of the atoms of which all materials are made. An atom consists of a nucleus, which is a positively charged particle that carries most of the mass of the atom. The electrons are much smaller and are each negatively charged. The total electric charge of all the electrons exactly balances the positive charge of the nucleus. Atoms are bound to each other by exchanging or sharing electrons with each other. When the electrons are exchanged, the binding is said to be *ionic*, when the electrons are shared, the binding is said to be *covalent*.

What materials are used as semiconductors?

The most common semiconductors are silicon and germanium, of which silicon is by far the more important. Other materials,

2

such as the compound materials gallium arsenide and indium phosphide, are also used.

How does silicon work as a semiconductor?

Silicon is a semi-metallic element and, like all semiconductors, it is crystalline, with its atoms arranged in a definite pattern, called the crystal lattice. Because of this structure, the crystal can be cut easily in certain directions, like that well-known crystal, diamond. A two-dimensional representation of a perfect crystal lattice is shown in Fig. 2.

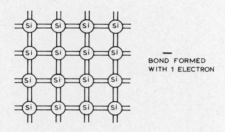

BOND FORMED
WITH 1 ELECTRON

Fig. 2. Representation of a silicon crystal lattice at absolute zero temperature, showing covalent (two-electron) bonds between adjacent atoms

The silicon atom has 14 electrons. Four of these are at a greater distance from the nucleus than the others and are available for forming bonds to other atoms (silicon or other elements). These electrons are called the *valence electrons*, and are less tightly bound to the atom than the other electrons. This arrangement is not peculiar to semiconductors; it is the normal structure of any element, but the number of valence electrons varies. All of the important semiconductor elements have four valence electrons.

In the crystalline structure of silicon, these valence electrons form *covalent bonds*, as shown in Fig. 2, with the valence

3

electrons of adjacent atoms. Each atom is equidistant from the four surrounding atoms, and each valence electron forms a covalent bond with one from an adjacent atom. The bonding is formed by the continual exchange of the two valence electrons which are involved.

In such a perfect crystal structure, imperfections can be formed when an electron breaks away from a covalent bond, carrying its negative charge with it, and also when one of the silicon atoms in the crystal is replaced with an atom of another element (a process which is called 'doping'). In a perfect crystal at the temperature called ' absolute zero', which is −273°C, all of the covalent bonds are complete, but at higher temperatures the energy of the electrons increases, and the exchange process is

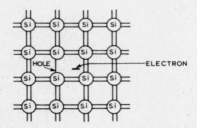

Fig. 3. The silicon crystal lattice at a higher temperature, with one free electron and one free hole in the structure

much more rapid. At such higher temperatures, a few electrons may have enough energy to break free. The gaps which are left in the covalent bonds are called *holes*, and behave in every way as charged particles. Unlike electrons, however, holes have no independent existence outside the crystal. This situation is illustrated in Fig. 3. An electron carries a unit of negative charge, while a hole, being a lack of a negative charge, represents an equal unit quantity of positive charge.

4

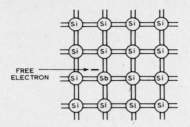

Fig. 4. A silicon crystal lattice in which one silicon atom has been replaced by an atom of antimony (symbol Sb). One of the antimony valence electrons is unused, and breaks free

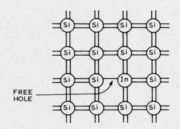

Fig. 5. A silicon crystal lattice in which one silicon atom has been replaced by an atom of indium (symbol In). All of the indium valence electrons are used, and there is one lacking, causing a hole to appear

Pure silicon at room temperature (20°C) has equal small numbers of holes and electrons, but impurities can be introduced by doping in order to modify this state. A suitable doping element will have an atom whose size is roughly the same as that of silicon so that it will fit easily into the crystal lattice, but which carries a quite different number of electrons. Correctly controlled doping can greatly increase the electrical conductivity of the material without disturbing the arrangement of the crystal lattice.

If an atom with five valence electrons (such as antimony) is used to dope silicon, a free electron is made available at every point where an antimony atom has been placed. This electron is not bound to any atom and can therefore pass freely through the material, taking its charge with it. The atom of impurity is known as a *donor*; it has donated an electron which can be used to carry electric current. The impurity is called a *donor impurity*, and the resulting semiconductor is said to be n-type, n meaning negative, since a negative particle, the electron, has been released.

An atom with three valence electrons, for example indium, may be used to dope the lattice, and in this case a hole will be left around each indium atom. As an electron fills a hole the hole appears to move around the crystal carrying charge with it, in this case positive charge. A semiconductor crystal which has been doped in this way will also conduct electricity well. The doping material is called an acceptor (of electrons) and the semiconductor is now p-type (p meaning positive) because the 'free particles' which have been released are holes.

In addition to these particles which are released by controlled doping, electrons are released by the action of heat or light, so that the conductivity of semiconductor materials is also affected by the temperature of the material and the intensity of light striking it.

What is a junction?

A semiconductor junction is a region where oppositely doped portions of the same crystal meet; the phrase *pn junction* is often used. When such a junction is formed, electrons on one side combine with holes on the other to form a 'no-man's land' a region which has very few particles capable of moving and carrying current. This region is called the *depletion region*, and it forms a barrier to the movement of current, which can, however, be modified by applying different voltages to the two sides of the junction.

What is a rectifying diode?

If an alternating voltage is applied to a junction of p-type and n-type semiconductor, the barrier presented by the junction to the flow of charge carriers between the two parts will be alternately strengthened and weakened. In other words, the resistance of the junction to current flow depends on the value and polarity of the voltage that is applied to it. Thus a rectifying action is obtained, and a diode based on this is called a junction diode or rectifier.

What is a point-contact diode?

The point-contact diode, usually formed using germanium, is still used for some applications. A very fine point of wire, made from a material which is a donor or an acceptor, is placed in contact with a doped crystal, and a current is passed. The wire welds to the surface of the semiconductor, and forms a very small, point-size, junction. Because of the very small size of the junction formed in this way, point-contact diodes have very small values of stray capacitance but can carry only small amounts of current. The small stray capacitance values make such diodes valuable for signal demodulation (detection).

What is meant by forward and reverse biasing?

If a d.c. voltage is applied across a pn junction–see Fig. 6 (a)–negative to the p section and positive to the n section, the junction is said to be reverse biased, the flow of charge carriers across the junction being retarded. On the other hand if the voltage is reversed in polarity, with negative to the n section and positive to the p section, Fig. 6 (b), the junction is said to be

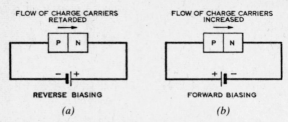

Fig. 6. Reverse (a) and forward (b) biasing of a semiconductor pn junction

forward biased, the flow of charge carriers across the junction then being increased.

What is the meaning of the expression 'solid-state'?

Solid-state is the term applied to diodes, transistors and indeed all semiconductor devices and the equipments that employ them. A solid-state device – as this chapter has indicated – depends on the interaction of electronic bonds in molecular structures, which of course are solid. In contrast, the electronic valve is a thermionic device in which free electrons move in a vacuum or through a gas.

Solid state is also used to distinguish purely electronic devices, using transistors and integrated circuits, from devices which use electromechanical components such as relays.

How is germanium obtained?

Germanium is obtained from two very different sources. One is the flue dust produced by burning particular types of coal containing about 0.02 per cent of the element. Burning the coal is the first stage of extraction, and the flue dust contains as much as two per cent by weight of germanium. The copper and zinc ores from parts of Africa form the other main source. The extracted material is sent in the form of germanium dioxide for transistor manufacture.

How is silicon obtained?

Silicon is one of the most common materials in the Earth's crust; sand is silicon oxide, containing about 30 per cent silicon by weight. Extraction is carried out by heating sand or other silicon-bearing mateial, mixed with powdered carbon, in an atmosphere of hydrogen. The silicon which is obtained in this way is much too impure to use for semiconductors, and the purification is carried out first by conventional chemical methods and finally by physical methods. The physical process used is recrystallisation, using a technique called 'zone-melting' which gradually concentrates the impurities at one end of a long crystal. Very pure silicon can be obtained by zone-melting, containing only one part of impurity per ten thousand million.

What is the main advantage of silicon?

As already mentioned, holes and free electrons will arise in semiconductor material due to the effect of heat, the heat providing sufficient energy to enable some electrons to break free from their covalent bonds. The resulting holes and free electrons give rise to unwanted currents which interfere with the desired flow of current in the semiconductor material. Since the 'energy gap' – the energy required by an electron for it to become free of its parent atom – is higher with silicon than with germanium, it follows that the operation of silicon semiconductor devices is less affected by heat than is the case with germanium devices.

A further advantage with silicon is the ease with which its surface can be oxidised, giving an insulating layer of silicon dioxide. This is of importance in the fabrication of most semiconductor devices, including integrated circuits.

Does germanium have any particular advantage?

While germanium is affected more by heat than silicon because of its lower energy gap, nevertheless this lower energy gap means that there is greater hole mobility in germanium. Because of this, germanium semiconductor devices are capable of higher frequency operation than silicon ones. It also means that they can be operated with lower supply voltages.

What is meant by 'majority' and 'minority' carriers?

As we have seen, two types of current can flow in semiconductor material. There is firstly that caused by the movement of electrons, which carry a negative electric charge (they flow towards the positive terminal of a battery) and thus give rise to a negative current; and secondly the movement of holes, which represents a positive current. By doping, an excess of electrons over holes or vice versa can be created. In the case of p-type material, doped with an acceptor impurity, there is an excess of holes, which are therefore called the majority carriers in such material. With n-type material, doped with donor impurity, there is an excess of electrons, and in this case the electrons are call the majority carriers.

When a pn junction is forward biased majority carriers will diffuse across the junction. Once across the junction they will be minority carriers, forming a flow of minority carrier current. The action of semiconductor devices is largely based on the effects of biasing pn junctions so as to control the currents flowing across them.

What is meant by the junction depletion region?

When a pn junction is first formed there is an initial flow of carriers across the junction – electrons in the n region are attracted across the junction by the positive charge represented

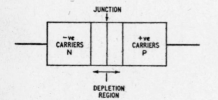

Fig. 7. Depletion region at the junction of a pn semi-conductor device. Few current carriers are present in the depletion region

10

by the holes in the p region, and vice versa. This results, as shown in Fig. 7, in a region on each side of the junction relatively sparse in charge carriers. This is called the depletion region. Once formed, the depletion region acts against further diffusion since there is no longer the initial field attracting holes and electrons across the junction. Forward biasing reduces or neutralises the depletion region; reverse biasing reinforces it.

What is a p-n junction characteristic curve?

A characteristic curve shows the electrical characteristics of a pn junction. A example is shown in Fig. 8. As can be seen, increase in forward voltage – forward bias – results in increased current

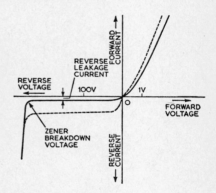

Fig. 8. Characteristic curve showing the electrical characteristics of a pn junction. The broken curve shows the effect of increased temperature on the characteristics. The reverse leakage current is mainly caused by holes and free electrons in the depletion region due to ionisation

flow. Reverse bias does not produce current flow until a critical point, the zener breakdown voltage, is reached, when a sudden substantial current which may destroy the junction flows. Prior to this point a small reverse current, called the reverse leakage current, flows across the junction. This is caused by ionisation – the effect of heat. The dotted curve indicates the change in characteristics – increase in reverse leakage current and forward current – brought about by increase in temperature.

11

What is meant by bipolar, unipolar and MOS transistors?

Transistors whose operation depends on using two pn junctions are called bipolar. Some devices, notably unijunctions and junction field-effect transistors, use one junction only, and are termed unipolar. MOS field-effect transistors have no junctions at all. More details of these important f.e.t. devices are given later.

2

THE TRANSISTOR

How does the transistor differ from the diode?

The transistor is a more elaborate device: it involves two
junctions arranged back to back (see Fig. 9). The first transistors

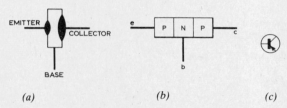

(a)   (b)   (c)

*Fig. 9. (a) A simple form of pnp junction transistor. (b)
Equivalent block diagram and (c) circuit symbol*

were of point-contact construction, but these have been super-
seded by the junction transistor, which is the main type to be
discussed here.

Are semiconductor devices a recent invention?

Crystal detectors were in use before the electronic valve was
invented (1904), and it is well known that galena and other
crystals were used by experimenters in the early days of radio.
However, further development of such devices had to wait for
many years.

In more recent times there has been renewed activity, with
contributions from specialists working in many scientific fields.

13

Among them were three research workers at the Bell Telephone Laboratories, in the USA, who in 1948 invented a three-electrode crystal device which could amplify as well as detect electrical signals. Like the early detectors, the new device used 'cat's whiskers' – two in this instance. The name 'transistor', decided on after study of its action, was derived from two words – transfer and resistor.

What is the action of a transistor?

As already indicated, the junction transistor consists of three 'layers' of germanium or silicon: these can be in pnp or npn form. There are three connections – to the emitter, the base and the collector. The base region (n-type in the pnp arrangement) is very thin and is lightly doped to give electron charge carriers. The regions on either side are heavily doped to give positive holes. The pnp arrangement is shown again in Fig. 10.

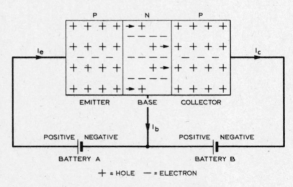

Fig. 10. A biased pnp transistor

Taking the base region as the reference point, the pn junction can be called the 'emitter junction' and the np junction the 'collector junction'. In Fig. 10 the emitter junction is biased by battery A in the *forward direction* (emitter positive with respect

14

to base) and the collector junction is *reverse biased* (collector negative with respect to base) by battery B. The p-region is more heavily doped than the n-region: thus most of the current that will flow across the emitter junction is due to holes from the p-region.

In addition, a current of electrons flows across the junction from base to emitter, but this is slight because of the heavy doping mentioned above. Forward biasing the emitter junction means that holes will be injected from the emitter into the base region. If a greater negative bias – from battery B – is applied to the collector, most of these holes will be attracted across the collector junction, through the collector region and towards the collector external connection. Here they will represent a positive charge which attracts electrons from the supply into the collector. What happens, therefore, is that holes flow through the transistor in one direction whilst electrons flow through it in the opposite direction, thus establishing a flow of current through the transistor.

In passing through the base region, some of the holes – typically less than one per cent – will recombine with free electrons in the base region, representing a flow of base current (I_b). This means that the flow of electrons to the collector should be less than one per cent different from the flow of holes to the emitter.

The important point to note, however, is that more than 99 per cent of the holes which cross the junctions pass from a low-resistance circuit (the forward biased emitter-base circuit) to a high-resistance circuit (the reverse biased collector-base circuit) – hence the term 'transfer-resistor', from which the word transistor stems. Thus although the current at the collector is less than that at the emitter, power amplification has been achieved (since power in Watts = $I^2 \times R$).

Is the npn transistor fundamentally different?

It is different in that the semiconductor sandwich form is reversed, as shown in Fig. 11, and the polarities of the applied voltages are reversed. Electrons, instead of holes, are injected at the emitter and collected at the collector, then flowing around the external circuit. Electrical characteristics are a little different from those of the pnp transistor described earlier.

Most silicon transistors are npn, because this type is easier to construct using silicon technology, but a wide range of silicon

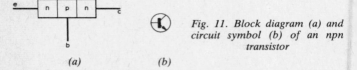

(a)   *(b)*

Fig. 11. Block diagram (a) and circuit symbol (b) of an npn transistor

pnp transistors is also available. The construction of germanium npn transistors was never so easy as the construction of the pnp type.

How can the transistor be made to amplify current?

The arrangement just described – called the common base configuration since the base is common to the emitter and collector circuits – provides, as we saw, power amplification. It also provides voltage amplification if a suitable load resistor is incorporated in the output circuit. The current gain, however, is less than unity. Suppose the biasing is rearranged as shown in Fig. 12, so that the emitter is the common connection. In this

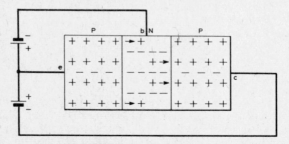

Fig. 12. Transistor with common-emitter connection

16

case the input consists of a small current flowing into the base region. Only a small current may be applied to the base, while the collector current will be relatively large. Since in an electronic device the gain is the ratio of the input to the output, current gain has been achieved with this configuration. The symbols β or h_{fe} are used to denote common-emitter current gain.

How is a bipolar transistor constructed?

In the comparatively brief time for which transistors have been manufactured, several processes have been used. Each successive process has resulted in better transistors, with greater current gain figures and less leakage. The process which is used nowadays to the virtual exclusion of all others is the *epitaxial diffusion process*. The problems of transistor manufacture are all concerned with base thickness and collector dissipation.

The thickness of the base layer of a transistor determines the amount of current amplification (h_{fe}) which can be obtained, and also the maximum frequency which can be amplified. Very much thinner base layers can be reliably obtained using the epitaxial diffusion process than were possible using earlier methods. The other problem concerns the fact that the movement of carriers through the material of the transistor causes heat to be dissipated. This heat must be conducted away through the collector layer (since most of the heat is liberated at the collector junction), so that a fairly thin, large area, highly conductive collector region is an advantage. The epitaxial diffusion process enables both of these problems to be more satisfactorily solved than did previous processes.

What is meant by epitaxial diffusion?

Diffusion means the gradual mixing of one material with another. If a bottle of perfume, for example, is opened at one end of a large room, its fragrance can soon be detected at the other end of the room. In this case, the perfume vapour has diffused through the air. Liquids diffuse into each other much more slowly (see Fig. 13), and solids hardly at all. A vapour can,

17

however, diffuse into a solid at a rate which can be controlled quite precisely by adjusting the temperature of the materials, and this fact enables us to form junctions on the surface of silicon by diffusing doping agents from a hot vapour.

12·00 NOON   5·00 P.M.

Fig. 13. Liquid diffusion – two liquids will mix of their own accord if they are given enough time

Epitaxy means the formation of crystals from a vapour of the same material. If a crystal of any material is in contact with vapour of the same material, the vapour will condense on to the solid, causing the crystal to grow in the same pattern of lattice as it had before. Both epitaxy and diffusion are used in the manufacture of silicon transistors and integrated circuits.

How is epitaxy used?

A silicon transistor starts as a very heavily doped crystal, called the *substrate* (meaning the layer underneath). The reason for having this layer heavily doped is so that it will be a good conductor of both heat and electric current, since all the collector current will have to pass through this substrate, and the dissipated heat will also have to pass through it. The much more lightly doped material which will form the collector region of the transistor is then grown on top of the substrate by epitaxy – bringing the substrate crystal into contact with hot silicon vapour which has been lightly doped. Air must be excluded from this process, because silicon oxidises readily, and the presence of silicon oxide prevents both epitaxy and diffusion on the oxidised surfaces.

18

How are the base and emitter layers formed?

Oxidation and photoetching methods are used. The substrate, with its more lightly doped collector layer on top, is oxidised in air, so that the collector layer is coated with silicon oxide. This layer also protects the purity of the material underneath. The oxidised surface is then coated with a photoresist – a material which hardens on exposure to ultra-violet light – and the coated surface is then exposed to ultra-violet through a mask which defines the shape and size of the base area. Where the photoresist has been exposed, the material has hardened, but the remaining photoresist can be washed away using hot water. By this means, the surface which will become the base region can be exposed.

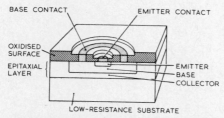

Fig. 14. Basic features of the silicon epitaxial planar transistor. In manufacture, the various regions are diffused through 'windows' etched in the oxidised surface, the surface being reoxidised after each process. The type shown is a small-signal a.f. transistor

The oxide layer is now removed from this exposed surface by means of an acid wash, and the material is dried and placed in a vacuum oven for a diffusion process. Silicon containing a doping material is used in vapour form, and the doping material is the opposite of that used for the collector region. The diffusion of this material reverses the doping of the part of the surface which it affects, and this is now the base region of the transistor. The oxidation and photoetching procedure is now repeated to form the emitter region, and this is diffused using the same doping material as was used for the collector.

19

Once again, the whole surface is oxidised, and a small hole is etched through to enable a metal contact to be made to the emitter. A similar hole and contact is made to the base region, so completing the construction. The complete piece of crystal, or chip, is now mounted in its container, metal or plastic, and the leadout wires are welded in place.

What is the field-effect transistor?

The f.e.t. is a type of transistor which does not depend on the action of two junctions. The main current flow is through a 'channel' which is made of material of one polarity, p or n, so that f.e.t.s are identified as p-channel or n-channel respectively. The two main types of f.e.t. are the junction f.e.t. and the m.o.s.f.e.t. (Metal-Oxide-Silicon f.e.t.).

What is a junction f.e.t.?

Fig. 15 shows the construction of a simple junction f.e.t. with a p-channel. Because there is a junction to a piece of n-type material constructed at one end of the channel, the cross-sectional area of the channel is very much smaller in the region around the junction, and a change of bias voltage on the n-type material (called the gate) will cause a considerable change in the number of carriers in this region.

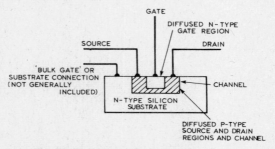

Fig. 15. Cross-section of a junction f.e.t.

The change is caused by the electric field which exists when a voltage is applied to the gate. As we have mentioned, there is a depletion region around any junction. When the junction is reverse-biased, the size of the depletion region increases, and this will make more of the area of the channel unusable as a path for current. Because of this, the resistance between the ends of the channel (the ends are known as source and drain respectively) is much greater, and less current flows through the channel.

Junction field effect transistors can be operated with the junction forward biased, but this would destroy one of the considerable advantages of the junction f.e.t. as compared to the bipolar transistor, that the current needed at the gate is negligible. For this reason, the junction f.e.t. is almost always

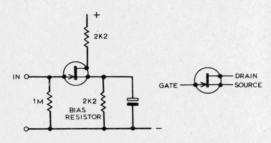

Fig. 16. How the gate of the junction f.e.t. can be reverse biased

operated with the gate reverse biased with respect to the source. Fig. 16 shows how the bias voltage for a typical junction f.e.t. (a 2N3119) is arranged.

What is a m.o.s.f.e.t.?

The m.o.s.f.e.t. uses no junctions, but the operating principle is similar to that of the junction f.e.t. inasmuch as the channel

21

conductivity is altered by using an electric field. The channel is shaped so that it has a very small cross-sectional area at one part. Over this part a thin layer of silicon oxide is grown and a layer of metal is placed on top of the silicon oxide by evaporation. Since the silicon oxide is an insulator, this arrangement constitutes a miniature capacitor, and the electric field which is caused by applying a voltage between the metal contact (the gate) and the channel will affect the number of carriers which are free to move in the restricted region of the channel.

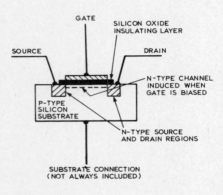

Fig. 17. Cross-section of a m.o.s.f.e.t.

Unlike the junction f.e.t., the gate of the m.o.s. type is non-conducting for either polarity of bias, and current never flows between the gate and the channel. M.o.s.f.e.t.s can be used in either depletion or enhancement mode. In the depletion mode of operation, the bias on the gate causes the channel to be depleted of carriers, so that the current between the source and the drain is reduced. In the enhancement mode, the bias on the gate causes additional carriers to be released into the channel, so that the resistance of the channel decreases, and the current passing between source and drain can increase. No current flows to or from the gate in either mode of operation.

What are the advantages and disadvantages of m.o.s.f.e.t.s?

The m.o.s.f.e.t. is entirely controlled by the voltage between the gate and the source, with no gate current flowing. This makes the device extremely useful when a very high input impedance is needed. The voltage gain, however, is rather low. Recently, power m.o.s.f.e.t.s (vertical m.o.s.f.e.t.s) have been developed which have considerable advantages for power audio output stages as compared to conventional bipolar output transistors.

What about handling problems?

The very high input impedance of the gate causes handling problems which affect all m.o.s. devices, transistors or i.c.s, because a voltage of more than about 50 V between the gate and the source will cause the thin silicon oxide layer to break down permanently. By contrast, the electrostatic voltage which can be developed by rubbing two materials together can easily reach several thousand volts. Bipolar transistors or junction f.e.t.s are not damaged by these voltages because the comparatively small amount of electric charge which causes the voltage can leak harmlessly through the junction, reducing the voltage to zero.

M.o.s.f.e.t.s are protected by keeping the gate and source electrodes shorted, either by metal contacts or by embedding the leads in conducting plastic until the devices are safely soldered into their circuit. M.o.s.f.e.t.s should never be soldered into a circuit until all other components have been soldered in place.

Does the power transistor differ from those already described?

Apart from the fact that it is a good deal larger, the power transistor is similar in essential respects to the other types described. It is however provided with a metal housing and, in

23

use, this is arranged to be in contact (via a mica insulating washer) with the chassis on which it is mounted so as to assist the dissipation of the considerable heat produced at the collector. If this precaution were not taken, the transistor would soon be irreparably damaged under normal operating conditions. A

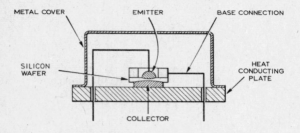

Fig. 18. Simplified cross-section of a power transistor

simplified section of a power transistor is shown in Fig. 18. It is of the kind that would be used in the output stage of a solid-state radio receiver.

3

BASIC TRANSISTOR CIRCUITS

How is a transistor connected into circuit?

Since the transistor is a three-electrode device, it can be connected in three ways, or configurations as they are called. These are the common-base, common-emitter and common-collector connections shown in Fig. 19. The term 'grounded' is sometimes used instead of 'common', since the common termination will often be taken to earth or chassis.

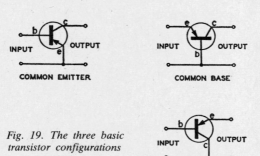

Fig. 19. The three basic transistor configurations

The main characteristics of these configurations are shown in Table 1. Exact values depend on the type of transistor and the nature of the load applied.

The common-emitter connection is the most frequently used arrangement for amplifiers; a voltage gain of 100 or more can be obtained using a single transistor. The common-collector connection is of interest in that, with its high input impedance and

Table 1. CHARACTERISTICS OF BASIC TRANSISTOR CONFIGURATIONS

Characteristic	Common emitter	Common base	Common collector
Input impedance	Medium	Low	High
Output impedance	Medium	High	Low
Current gain	High	Less than unity	High
Voltage gain	High	High	Less than unity
Phase inversion	Yes	No	No

low output impedance, it corresponds to the cathode-follower circuit used in valve amplifiers. The arrangement is used for a matching stage between transistors in common-emitter connection. The common-base arrangement finds its main application to v.h.f. and u.h.f. amplifiers and oscillators.

What is phase inversion, referred to in the table?

It is noted in the table that, in the common-base and common-collector configurations, there is no phase inversion of the signal. In other words, the input and output signals are in step: they both swing positive or negative at the same instant (see Fig. 20). In the common-emitter configuration, however, the output signal swings to positive when the input signal is negative. Thus there is phase inversion, which as far as a sine wave is concerned, is the same as the waveform produced by a 180° phase shift (pushing the wave forwards or backwards by half

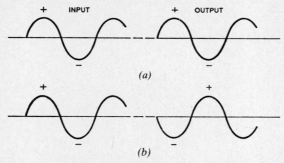

Fig. 20. (a) Input and output signals in phase; (b) phase inversion: output signal is 180° out of phase with input

a cycle). For waves of other shapes however, phase inversion does not produce the same waveform as a 180° phase shift.

How is a transistor biased?

We have seen in Figs. 10 and 12 the biasing arrangements required for common-base and common-emitter transistor stages, using two batteries. The use of two batteries, however, is not a practical arrangement. A simple method of biasing a common-emitter stage using a single battery is shown in Fig. 21(a). Here a small current from the supply flows in the base circuit as a result of connecting the bias resistor R_{bias} between the base and the positive supply line. This will make the base positive with respect to the emitter, the situation we require, with the npn configuration shown, to forward bias the emitter junction.

In practice more elaborate biasing arrangements are generally used in order to overcome the increase in current flow with rise in temperature that occurs in semiconductor material. As a transistor heats up when operating, so the current flowing through it will increase; and the greater current flow will cause more heating, and so on. When a resistive load in the collector or emitter circuit limits the amount of current which can pass, this effect will not damage the transistor, but will lead to severe distortion of amplified signals. In low-resistance circuits, such as power output stages, the current is not limited, and the condition

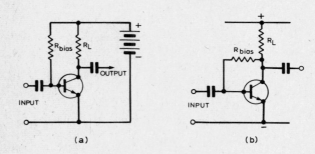

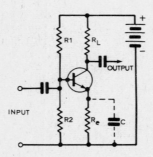

Fig. 21. Methods of biasing a transistor. NPN transistors are shown in these examples

called thermal runaway can occur. In this condition, the temperature and current increase steadily until the junctions are damaged and the transistor destroyed.

Two methods of overcoming the problem are shown in Fig. 21. In Fig. 21(b), a resistor is connected between the collector and the base terminals, and the current flowing in this resistor is the base bias current. Since an increase of collector current will cause the collector voltage of such a circuit to drop, the bias is automatically reduced. This method is more suited to a circuit with a resistive load in the collector circuit, and for low resistance collector loads, the circuit of Fig. 21(c) is more suitable.

A potential divider (R1, R2) is used to provide a stable bias, and a small resistor R_e is added in the emitter lead. The effect of R_e is that as the emitter current rises because of heat so the base voltage with respect to the emitter (with an npn transistor) falls, thereby pulling back the base current and the collector current. As shown, with an a.c. signal a decoupling capacitor C may be

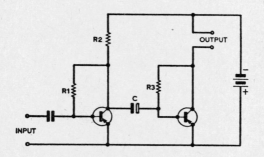

Fig. 22. A simple two-stage transistor amplifier circuit. It is important in using electrolytic capacitors in transistor circuits to observe correct polarity. Generally with pnp transistors in cascade the collector of the first will be more negative than the base of the second, so that the electrolytic coupling capacitor is connected as shown here

29

included to overcome the negative feedback that would otherwise be introduced through voltage variations at signal frequency
across R_e.

Describe a two-stage amplifier circuit

A simple two-stage amplifier circuit diagram is shown in Fig. 22.
This circuit uses the bias method of Fig. 21(b), with a resistor
connected between the collector and the base of each transistor.
Resistance-capacitance coupling between the two stages is used,
so that the output signal voltage from the first transistor is fed
through capacitor C to the base of the second stage. Because of
the low resistance of the base circuit of a transistor, large-value
electrolytic capacitors have to be used, otherwise low frequencies are attenuated. These electrolytics must be connected so
that the correct polarity is observed.

Fig. 23 shows a direct-coupled amplifier pair which uses two
npn transistors. The second transistor is biased by the collector
voltage of TR1, and the base of TR1 is biased by the voltage
across the emitter load of TR2, R_{E2}. This bias arrangement is a

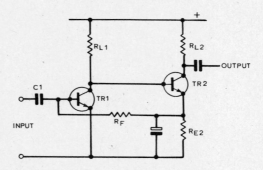

*Fig. 23. A two-stage direct-coupled amplifier. The bias resistor R_F is
fed from the emitter of TR2, so that the bias of both transistors is
controlled by the same resistor*

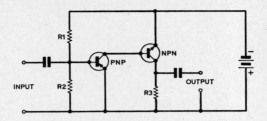

Fig. 24. A complementary circuit, using pnp and npn transistors, gives direct coupling between stages.

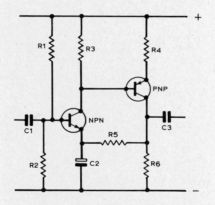

Fig. 25. A complementary amplifier which uses feedback bias through R5. This arrangement also combines high gain with good stability. Note that the order of using npn and pnp has been reversed as compared with Fig. 24

31

more elaborate version of the scheme of Fig. 21(b), and is called feedback biasing. It results in very stable bias conditions, because any changes in current and gain are self-correcting.

A complementary arrangement is often useful. The word complementary is used to describe circuits which make use of both pnp and npn transistors, or both p-channel and n-channel f.e.t.s. The example shown in Fig. 24 is of a two-stage amplifier which makes use of pnp and npn transistors so that there is direct coupling between the stages. Such an arrangement uses very few components, has a high gain and good low-frequency response (because no coupling capacitor is needed between the transistors), and can be biased so that the d.c. level (with no signal applied) at the output is almost the same as that at the input. The output and input signals are in phase. Fig. 25 shows how some d.c. feedback can be added to make the bias conditions more stable.

How is the transistor used in oscillator circuits?

A simple r.f. oscillator is shown in Fig. 26. The base is biased on by the potential divider network (R1, R2) so that current flows

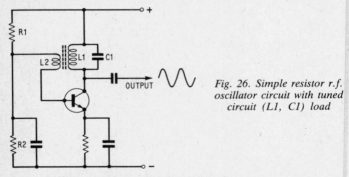

Fig. 26. Simple resistor r.f. oscillator circuit with tuned circuit (L1, C1) load

through the transistor, whose load is the tuned circuit L1, C1. Part of the output is fed back to the base via the small coupling winding L2, resulting in sustained oscillation provided that the

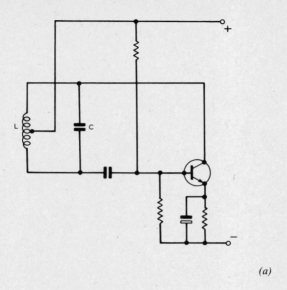

(a)

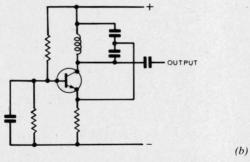

(b)

Fig. 27. The Hartley (a) and Colpitts (b) radio frequency
oscillator circuits

33

winding is arranged with the correct phase relationship to ensure that the feedback is positive.

Fig. 27 shows two other sine wave oscillators, the Hartley and the Colpitts. The principles are the same: a small portion of the waveform at the output is fed back in phase to an input, so that oscillation continues at the frequency determined by the inductance and capacitance values. Fig. 26 shows an important form of untuned oscillator, the astable multivibrator. The two amplifiers are connected so that an input signal which is inverted

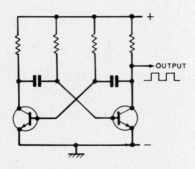

Fig. 28. The astable multivibrator. The high gain of the two-transistor arrangement ensures rapid switching, which causes the steep sides of the waveform

by one transistor is further amplified and inverted again by the second, and then fed back to the first. Because of the very great amount of amplification, the output is not a sine wave, but a steep sided rectangular wave (Fig. 28), and its frequency is decided by the values of the coupling capacitors and the base bias resistors.

4

TRANSISTORS IN RADIO AND AUDIO

How extensively are transistors used in telecommunications?

It is taken for granted that all radio receivers are designed around transistors or i.c.s. In the UK, all TV receivers are now transistor operated, and contain several i.c.s also, though this is by no means true in the USA. The higher-power radio transmitters are valve equipped, but only where valves are essential, which is in the final stages of power amplification. Signals at lower levels (such as exist in studio equipment, for example) are invariably handled by transistors.

Aren't valves making a comeback in audio amplifiers?

Transistor amplifiers have developed to a stage where the distortion caused by the amplifier is negligible compared to that caused by pickups, tuners, loudspeakers or even connectors! Many audio enthusiasts who have never heard valve-operated amplifiers are struck by the difference – which is due mainly to higher, not lower, distortion levels. A valve 'comeback' is therefore announced at intervals, but the steady trend over the last 20 years of audio is towards grouping equipment into fewer and smaller boxes, which precludes the general use of valves.

What about v.f.e.t.s?

Higher power f.e.t.s, described as vertical f.e.t.s (v.f.e.t.s) because of their internal construction, have appeared in several audio amplifiers at the upper end of the price range. These appear to offer considerable advantages as compared to conventional bipolar transistors in audio output stages, and more extensive use of v.f.e.t.s can be expected in the future.

What is the transistor's function as amplifier in a receiver?

The basic conditions for amplification have already been briefly dealt with. Next it is useful to consider amplifiers in relation to radio receivers and, later, more specialised kinds of sound reproducing equipment.

In the final stages of the receiver the problem is to obtain the maximum output safely and with minimum distortion of the signal, or at any event to achieve a level of distortion that will be acceptable to the listener. Several modes or classes of operation of amplifier stages are possible.

What are these classes of operation?

Three classes are important in receiver output stages and in more specialised audio work (to which reference is made later); or, to be more precise, two main classes, class A and class B, and a

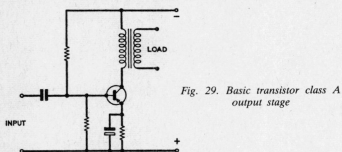

Fig. 29. Basic transistor class A output stage

sub-class AB. This form of classification relates to the d.c. bias applied to the transistor. The following is a brief outline of the subject.

The essentials of a class A transistor output stage are shown in Fig. 29. To obtain the greatest efficiency the load and input signal must be so adjusted that the collector voltage and current vary between zero and twice the no-signal value. The *average* collector current is the same whether or not a signal is being handled: thus the power taken from the supply remains constant. This means that the power dissipated *in* the transistor is reduced when the signal is applied. The theoretical efficiency of the class A stage is 50 per cent (that is, under ideal conditions); in practice it is a somewhat lower, variable figure.

In class B operation the 'quiescent' current – that flowing without application of a signal – is very small, but the current increases as the amplitude of the input signal increases. Since the drain on the supply depends in this way on the signal, economies are possible. The theoretical efficiency is 78.5 per cent. However, the output waveform from a simple class B stage is a series of half-cycle pulses, and it is therefore necessary to use two transistors in push-pull in order to obtain a usable waveform (see Fig. 30). This arrangement offers a relatively high power

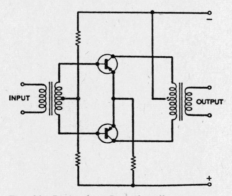

Fig. 30. Basic class B push-pull output stage

handling capacity but very careful design is necessary to avoid audible distortion.

The circuit of Fig. 30 is still used in miniature radios and portable cassette recorders, but it is much more common now to

37

use output stages which have no transformers in the signal path. Such an output stage, called a 'totem-pole' circuit, is illustrated in Fig. 31. TR2 is npn and TR3 is pnp, with both bases driven by the signal at the collector of TR1. TR2 will conduct for the positive half-cycles of the output, and TR3 will conduct for the negative half-cycles, so that the term push-pull is as valid for this circuit as it is for the circuit using transformers. The diodes D1, D2 in the collector circuit of TR1 ensure that both TR2 and TR3

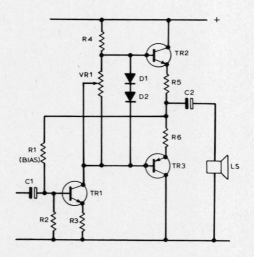

Fig. 31. A totem-pole output circuit, simplified. TR1 is a voltage amplifier whose load resistor is R4. The diodes D1, D2 ensure that there is a d.c. voltage difference between the bases of TR2 and TR3, which are driven by the same a.c. signal from the collector of TR1. Each of the output transistors then acts as an emitter-follower, passing signal through C2 to the loudspeaker

38

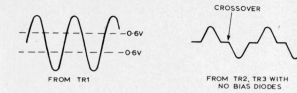

FROM TR1

CROSSOVER

FROM TR2, TR3 WITH
NO BIAS DIODES

Fig. 32. Crossover distortion. Unless there is a bias voltage between the bases of TR2 and TR3, approximately the first 0.6 V of the signal in each direction (positive or negative) will not be amplified. This produces a very noticeable form of distortion called crossover distortion

conduct when there is no input signal, they have no effect on the signal itself. If these diodes were not included, then the first 0.6 V of the signal, both above and below the steady bias voltage, would not be amplified (Fig. 32). This would cause severe distortion, a type called crossover distortion.

How much power is obtainable from transistor output stages?

As much as you need and can pay for! Small battery operated radios and cassette recorders use class B circuits with output powers of up to 2.5 W. For mains-powered audio equipment, only cost limits the power levels, and stereo amplifiers with outputs of 100 W per channel or more are available.

What types of oscillators are used in transistor circuits?

A few types of oscillator circuits have been briefly mentioned. There are two main classes of oscillators, the tuned oscillators which generate sine waves, and the aperiodic (untuned) oscillators which generate steep sided waveforms. Tuned oscillators, such as the Hartley and the Colpitts, use either *LC* (inductor-capacitor) tuned circuits or quartz crystals to determine the frequency of oscillation of the sine wave. Aperiodic oscillators, such as the astable multivibrator, use the transistor as a switch, conducting only very briefly in each cycle, and making use of the charging of capacitors through resistors to determine the frequency of the wave.

How are radio signals demodulated?

The radio signal consists of a symmetrical wave (Fig. 33) which can have no effect on a loudspeaker or earphone. Demodulation is the process of extracting the audio signal from this wave. A

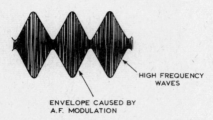

HIGH FREQUENCY
WAVES

ENVELOPE CAUSED BY
A.F. MODULATION

Fig. 33. A radio signal. The high-frequency carrier wave has its amplitude changed (modulated) by the audio signal, so that the audio signal shape forms an 'envelope' for the radio-frequency wave. No loudspeaker or earphone can reproduce the radio-frequency signal, however, and the envelope is symmetrical so that the audio modulation can have no effect on these devices

typical demodulator or detector circuit is shown in Fig. 34; it consists of a diode feeding a resistive load with a capacitor used to store charge. At each radio frequency cycle, the diode conducts whenever the voltage at the anode is greater than the voltage at its cathode, and so charges the capacitor to the peak of the radio frequency voltage. The resistive load is chosen so that the charge on the capacitor will change by only a small amount

40

between the radio frequency peaks. Since the amplitude of the r.f. *peaks* is changing comparatively slowly, the voltage across the capacitor can follow this change faithfully, and this variation constitutes the audio signal. The diode makes use of only half of the radio frequency wave, so removing the symmetry. A further filter stage to remove any traces of the radio frequency may be needed.

What is a.g.c. and how is it incorporated?

Automatic gain control is a regulating technique. Its purpose is to prevent the overloading of the later stages of the receiver that would otherwise occur with the reception of strong signals. On the other hand, where there is a fading of signals (common in a.m. reception), a.g.c. helps to minimise the effect on listening conditions.

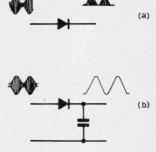

Fig. 34. A diode used by itself (a) will pass only one half of the radio signal. Adding a capacitor (b) will cause the output to consist of the envelope only, with little trace of the radio frequency

A control voltage is derived from the detector and applied to an i.f. stage, often in series with the base-emitter bias voltage of the first i.f. transistor. The control voltage is proportional to the carrier amplitude at the detector and controls the gain of the

41

stages to which it is applied. The arrangement is such that weak signals (i.e. fading) permit a high gain in the i.f. stage, whereas strong signals reduce the gain. It is necessary to include filtering components to eliminate audio and radio frequency signals from the a.g.c. line.

What is the i.f. amplifier stage?

As Fig. 35 shows, the i.f. stage is very similar to the simple amplifier stages already dealt with, but it has input and output

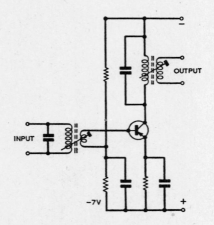

Fig. 35. Basic transistor i.f. amplifier stage

transformers which are tuned to the intermediate frequency – usually 470 kHz in a.m. receivers. These transformers determine the frequency range over which the stage provides amplification by admitting and passing on only signals within that range.

Audio transistors cannot be used for r.f. and i.f. amplification at high frequencies, because the capacitance between the collector and the base allows too much signal feedback, and the thickness of the base layer causes the transistor to lose gain at high r.f. frequencies. Transistors with thinner base layers and much lower capacitance values must be used, particularly for frequencies of 10 MHz and more. The usual i.f. of 470 kHz which is used for a.m. radios is so low that most transistors, apart from power types, of modern design will operate satisfactorily. At one time, circuit 'dodges' such as neutralisation and unilateralisation had to be used to compensate for the deficiencies of the early transistor types.

How is the transistor used in the mixer stage?

By means of a heterodyne (or 'beating') action the various frequencies of the received signals are converted to one fixed frequency, usually 470 kHz, at the input to the i.f. amplifier. The audio information is conveyed by the radio-frequency waveform as a modulation of amplitude in a.m. reception (or of frequency in f.m.), and this is transferred to the i.f. carrier. The intermediate frequency is above the audio range and can therefore be called supersonic: this combined with heterodyne many years ago gave us the name 'superhet', which is applied to receivers in general use today. The simpler 'tuned radio frequency' (t.r.f.) receiver is to all intents and purposes obsolete.

The functions of mixing and of oscillation have to be combined in some way at this stage in the receiver. Although two transistors and their associated components may be used, it is quite normal to arrange for one transistor to do all that is required. Careful design ensures that there is little difference in performance, and of course the cost can be reduced – a factor of great importance in the production of receivers of the popular type.

A stage of this kind is known as a self-oscillating mixer, and its essential features are shown in Fig. 36. The base circuit is tuned to accept the radio-frequency input, while the output circuit is tuned to the intermediate frequency output. The LC feedback

43

circuit between the collector and emitter results in oscillation, and the intermediate frequency is the difference frequency between the radio frequency input and the frequency of the stage's self-oscillation. It is important to note that the mixing action is carried out only because the transistor is being operated in a non-linear way (not as a linear amplifier).

A practical self-oscillating mixer circuit is shown in Fig. 37. For simplicity, this shows the components for one waveband only; for other bands, a switch (the wavechange switch) selects other coils and sometimes other capacitors in use in the circuit. To ensure that oscillation starts easily, the transistor is biased initially into Class A operation by the potential divider circuit at the base. As the oscillation builds up (a matter of one or two

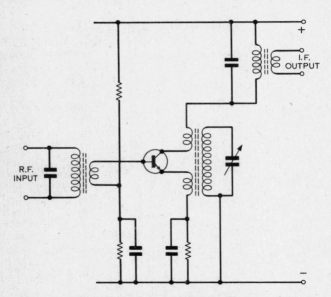

Fig. 36. Simple self-oscillating mixer stage. This type of circuit is often referred to as an 'additive' mixer

44

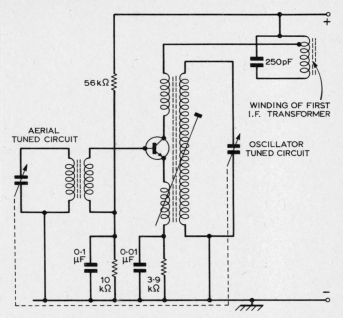

Fig. 37. Self-oscillating mixer stage for a pocket transistor radio

radio frequency cycles) the transistor starts to cut off in each cycle as the emitter voltage is taken positive; this is the non-linear action which causes the mixing. Measurements of the d.c. voltages of the electrodes can be very misleading, indicating that the transistor is non-conducting. It is the cut-off condition caused by the voltage swing at the emitter which stabilises the oscillator voltage.

How are these various stages combined?

The circuit of a five-transistor miniature receiver (Fig. 38) illustrates how the individual features which we have discussed

45

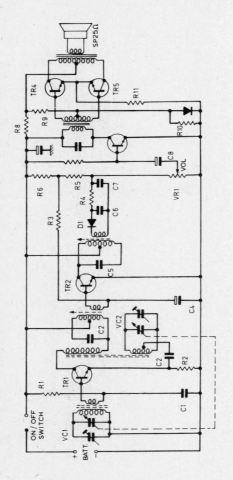

Fig. 38. Circuit of modern single waveband radio using silicon transistors

are combined. In this receiver, a ferrite rod is used as an aerial. The ferrite rod is made from material which is a very sensitive detector of changes of magnetic field, so that it can detect the magnetic portion of a radio signal just as the conventional wire aerial can detect the electric portion of the signal. The coil which is wound round the ferrite rod fulfils the dual function of converting the magnetic signals into voltage signals (using electromagnetic induction) and of tuning the incoming signal, using VC1 as the tuning capacitor.

TR1 is a mixer-oscillator of the type which has already been described, and TR2 is an i.f. amplifier. The amount of voltage gain (amplification) which can be obtained using modern silicon transistors at 470 kHz is so large that only one i.f. stage is needed, in contrast to early designs which normally used two such stages. A diode detector is used, with a.g.c. provided through R3, R6 to the base of the i.f. transistor. The output stage is the old-style transformer-coupled type, which can provide more power at the low supply voltage of this receiver (6V) than the transformerless designs.

How are printed circuit boards made?

Virtually all commercial electronics circuits are constructed on printed circuit boards (p.c.b.s), since these eliminate the cost and uncertainty of hand-wiring. The p.c.b. starts as a sheet of laminate, which may be synthetic resin bonded paper (s.r.b.p.) or glass fibre, coated with a layer of copper. The pattern of connections which will be needed is drawn on a large sheet of paper and carefully checked. The pattern is then photographically reduced to the correct size and transferred to a transparent sheet. By using a coating of photo-resist (see page 19) on the copper and exposing to ultra-violet light through the transparent master sheet, the resist can be hardened in the places where connections are desired, and the remaining photo-resist washed off. The exposed copper is then etched away, using a mildly acidic material such as ferric chloride which does not affect the hardened photo-resist.

Since the pattern has been protected by the photo-resist, the board can now be scrubbed clean, leaving the copper lines of the circuit surrounded by the insulating plastics board. The board is then drilled so that the component leads can be inserted. All the components can be soldered to the board in a single operation by dipping the copper side of the board with the components inserted into a bath of molten solder.

Fig. 39. A typical example of printed wiring. The black channels of conducting material form the 'wiring'. Component leads are inserted in the small holes shown, and soldered

Specialised boards use printed circuit patterns on both sides of the boards, with connections made from one side to the other by means of holes which are plated with copper around their sides (plated-through holes).

What are the advantages of using p.c.b.s?

The development of printed wiring has been paralleled by the miniaturisation of components, so that much lighter and more compact assemblies become possible. Speed of soldering has already been mentioned, and another factor is the reduction of sundries such as tags and eyelets, which are of course used in considerable numbers with conventional wiring.

Therefore faster and simpler assembly is possible: production rates are increased and costs can be reduced. Again, it is claimed that the incidence of electrically unsound joints (usually called 'dry' joints) is substantially reduced.

What about the use of i.c.s?

Curiously enough, i.c.s have hardly been used in portable radios, though they are extensively used in car cassette players and in hi-fi equipment. I.C.s are dealt with in more detail in Chapter 5.

Are transistors still used in TV receivers?

To a large extent, though modern receivers use i.c.s extensively. Transistors still dominate in the r.f., mixer and i.f. sections, though demodulators and the colour processing stages almost invariably make use of i.c.s nowadays, particularly in European and Japanese receivers. The line and field output stages makes use of transistors, because of the amount of power which has to be dissipated in these stages. Some modern designs now use an i.c. as the field output stage, however, with an identical type used as a sound output stage. Oddly enough, portable TV receivers, apart from the miniature types, make rather less use of i.c.s than the larger models. Receivers which feature *teletext* services (Ceefax, Oracle and Prestel) use large numbers of digital i.c.s for the decoding and storage of text, and remote control of TV receivers also needs digital i.c.s.

What sort of developments have taken place in high-fidelity amplifiers?

The advent of low-cost silicon transistors started a rapid trend to high-performance transistor amplifiers which has continued ever since. The 'totem-pole' type of output stage is practically universal, with minor variations, and the quality of performance relies on careful selection of components, attention to biasing, physical layout and other features which are time-consuming at the design and production stages, and which make a really good amplifier expensive.

A feature of recent work has been emphasis on transient performance, requiring amplifiers to be able to follow rapid voltage changes which are not found in recorded music. All but the cheapest (in the quality sense, which is not necessarily the cost sense) amplifiers offer performance levels which would at one time have been considered unobtainable, but the effect of making better equipment is that listeners become more critical, so that improvements are still being sought. One recent trend at the time of writing has been to introduce miniature amplifiers, with matching tuners, cassette decks and even record decks. These miniature units show little sacrifice in performance, but present a rather fussy appearance due to the use of so many small boxes piled on top of each other.

What types of output stages are used?

The simplest type of class B output stage was shown in Fig. 31. This requires complementary pnp and npn transistors, and it is not easy to obtain well-matched pairs of such transistors for

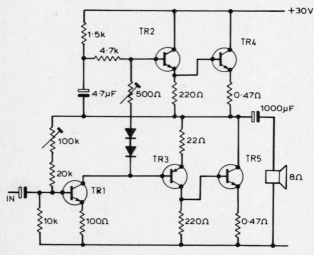

Fig. 40. Quasi-complementary symmetry output stage fairly typical of circuits used in cheaper types of hi-fi amplifiers

every range of power output. For some time, the quasi-complementary circuit illustrated in Fig. 40 was used as a solution to this problem. In the quasi-complementary circuit, the lower-power complementary pair of transistors drive the higher power npn output transistors, separating the action of phase splitting (done by TR2, TR3) and power amplification (TR4, TR5). The disadvantage of the circuit is that it is not truly symmetrical, causing the distortion levels to rise considerably as the output power is increased.

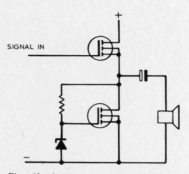

SIGNAL IN

Fig. 41. A v.m.o.s. output circuit, in which one v.m.o.s. transistor is driven with the signal, and the other is used to set bias

There has been a continual improvement in the quality of matched pnp/npn output transistors, however, in recent years, so that the quasi-complementary circuit is used less often than it was. A more recent development has been that of high power f.e.t.s, as mentioned in page 23. A typical circuit is shown in Fig. 41.

What is a current-dumping stage?

Current dumping is a term used by the Quad Acoustical Co. for a circuit method devised by their chief engineer and patented

51

worldwide. Briefly, the principle is to use a very linear class A amplifier as a main output stage. This amplifier is a low power one, but it controls, by use of feedback, a pair of high-power silicon transistors, called the current dumpers. The additional transistors are used to supply the extra current which is needed when large-amplitude signals have to be handled. Since they are completely controlled by the low-power class A amplifier, the distortion is due to this amplifier alone, and can be made very low indeed. At the time of writing, only the Quad Acoustical Co. has used the technique.

What of the other stages in an amplifier?

The preamplifier stages make use of modern silicon transistors in class A voltage amplifying circuits. The difference between a really good design and an inferior one is seldom apparent from the circuit diagram alone. It also depends on careful attention to bias levels, protection from overloading, selection of transistor types and characteristics, good specification of other components (such as low-noise resistors, for example) and good physical layout. In particular, it is important to note that two transistors of the same type, made on the same production line at the same time, can be vastly different in behaviour. The manufacturers of high quality equipment will therefore pay considerable attention to testing and selecting transistors since this can be more rewarding than continually re-designing circuits.

The use of i.c.s in high-quality equipment is almost negligible, as far as amplifiers are concerned, though tuners and lower-priced audio circuits make extensive use of i.c.s. The wide tolerances which are invariably found in i.c.s make them unacceptable for use in high-quality amplifiers, except in ancillary roles (controlling dial displays or meter readings).

How do tape recording circuits differ from amplifier circuits?

The main portion of the circuitry of a tape or cassette recorder consists of an amplifier whose voltage amplification stages and

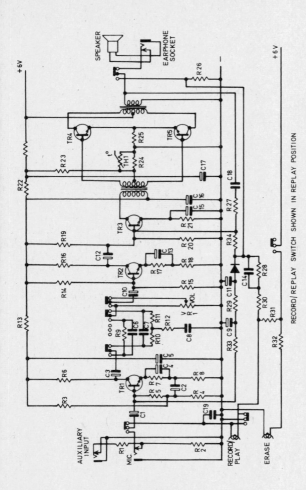

Fig. 42. A cassette recorder circuit typical of a small recorder. Many modern recorders are appearing with one i.c. replacing most of the circuit shown here

RECORD/REPLAY SWITCH SHOWN IN REPLAY POSITION

53

power output stage will be identical to those used in radios and in audio amplifiers. In addition however, a preamplifier is needed with a voltage gain considerably greater than the voltage amplifier stage of a radio. The typical signal level at the tape head is about 1 millivolt (1/1000 volt), as compared to the 1V or so which is available at the demodulator of a radio. The preamplifier must also act to correct the frequency distortions which are caused by the tape recording process, mainly a lack of treble.

A typical cassette recorder circuit is shown in Fig. 42. The record/replay switch is shown in the replay position, so that the replay action will be considered first. One side of the replay head is earthed, and the signal from the other terminal is taken to the base of TR1, a voltage amplifier. The amplified signal from the collector of TR1 passes through an 'equalising' circuit, using resistors and capacitors, to correct the distortions caused by the recording process. The values of the resistors and capacitors are chosen so that signals which had constant amplitude but varying frequencies when recorded will still have constant amplitude after passing through the equalising circuit, though the amplitudes would certainly not be equal at the collector of TR1. TR2 and TR3 act to amplify the equalised signal, and to drive the output stage, comprising TR4, TR5. This is a conventional transformer-coupled stage, used because of the low battery voltage (6V) which is employed. Some negative feedback from the loudspeaker output to the base of TR3 is used to reduce the distortion caused by the output stage. The erase head and the microphone and other inputs are inactive on playback, as is the network of resistors and capacitors to the left of R34.

On recording, the main record/replay switch is reversed, so that the other side of the replay head is now earthed and this head is used as a recording head. The base of TR1 is now supplied with the output from the microphone or the auxiliary input, and a different equalising circuit is switched into place between TR1 and TR2. TR2 and TR3 are used as voltage amplifiers as before, and the output stage is switched from the loudspeaker to a resistor. The signal across this resistor is used to drive the recording head, and is also rectified by D1 and used to

54

bias TR1. This bias on TR1 provides an automatic gain control, so that the volume control does not have to be used during recording. The erase head on this low-cost recorder is supplied with d.c. through R32.

Other designs ensure higher-quality recording by using a.c. supplied to the erase head and the recording head. In such designs, the output stage is connected as an oscillator during recording, making use of a more elaborate record/play switching arrangement.

Are special amplifiers used for public address?

The amplifiers which are used for public address systems are generally of similar design to those used in hi-fi. The main differences are that the supply voltage is usually 12V or 24V, to make use of car batteries, and the output stages incorporate transformers. The use of output transformers enables long lines to be driven. By using high signal voltages and low currents (obtained by using an output transformer), large amounts of audio power can be supplied to loudspeakers at a considerable distance, along long lines, without having an excessive loss of power due to the current flowing through the resistance of the cables. Another transformer at each loudspeaker then matches the high-voltage, low-current signal on the lines to the low-voltage, high-current signal which is needed by the loudspeaker.

What can cause damage to transistors?

Transistors, used correctly, have almost indefinite life, so that a correctly designed circuit should never permit a transistor to be damaged. Many of the early transistor radios are, for example, still operating perfectly. The breakdown of other components may cause damage to transistors by allowing excessive currents, particularly if the excessive current is between the base and the emitter. Another source of trouble is a regulated voltage supply which because of a regulator fault becomes unregulated, allowing excessive voltage to be applied to transistors. The third, and probably most common cause of trouble, is excessive temperature rise.

Silicon transistors are much less affected by temperature rise than the older germanium transistors, but damage is still possible if the transistor is being used near the limits of its safe operating ratings and there is an excessive temperature rise in addition.

What precautions are necessary with transistors?

As indicated above, it is essential to guard against incorrect electrical operating conditions and to avoid sources of heat that will raise the transistor's temperature above safe limits. The working temperatures of silicon transistors can be higher than those of germanium devices.

For power transistors, such as those used in audio output stages, it is necessary to provide a *heatsink*. This may take the form of a clip or collar which will dissipate heat by convection. In many cases the clip is joined to the amplifier chassis, which can of course conduct and dissipate a considerable amount of heat. The housings of some transistors are designed to be positively located on a chassis, making thermal contact with the metal. In other instances large fins are used to give the biggest possible radiating area.

Without a heatsink the transistor may suffer *thermal runaway*. Heat dissipated in the transistor junction raises the local temperature; this raises the leakage current, which in turn increases the dissipation even more. The limit is reached only when the transistor is destroyed by the heat.

Such matters are of course attended to at the design stage, but precautions that are similar in principle are necessary when equipment is repaired. In particular it is important to hold the lead of a transistor with pliers when it is being soldered into place or removed. The pliers, held between the soldering iron and the body of the transistor, act as a cooling fin. It is important to use an earthed iron: a few volts may exist at the bit relative to earth, and this potential may send a damaging current through the transistor.

Care is needed when a transistor is replaced in a heatsink. Insulating washers and bushes must be correctly replaced, and heatsink grease (a special silicone grease) must be smeared on all the surfaces which are to be clamped together. The clamping bolts should be done up tightly, so that the heatsink grease is squeezed out. This will ensure an unbroken film between the surfaces.

Again, measuring instruments must have a very small output voltage (a maximum of 1.5 V is advised by some manufacturers), but for some tests it is wise first to remove the transistor from circuit. Equipment under test should be switched off before components are changed, so as to rule out the possibility of current surges that would damage the transistors. A short-circuited junction is the type of fault caused by carelessness in this respect. An open-circuited junction is usually caused by overloading, possibly due to a fault elsewhere in the equipment.

With reasonable care mechanical damage should not occur. However, the terminal leads must not be strained or bent very close to the seal at the base of the housing.

5

OTHER SEMICONDUCTOR DEVICES

What other devices are in use?

The zener and varicap diodes are especially important; then there are the unijunction, tunnel diode, and phototransistors representing very different applications of semiconductors. Thyristors and triacs, along with diac diodes are used in power control, and l.e.d.s to produce light. A semiconductor material of a very different type is used in thermistors.

What is a tunnel diode?

This device, invented by Dr. Esaki of Japan, has not fulfilled its early promise, owing to the difficulties of using it in circuits. It is a microwave oscillator, but it has been replaced for most purposes by other specialised diodes such as IMPATT diodes.

The characteristics of the tunnel diode are due to heavy doping of the p and n regions. As a result of this there is a steady increase in reverse current with increase in reverse voltage while in the forward direction there is a sharp increase in current at first with low forward voltage applied. This initial current is called the tunnelling current and after about 0.15 V gradually dies away to be followed at about 0.3 V by the normal flow of forward current. The negative resistance characteristic is due to the changeover between these two current flow mechanisms in the forward direction.

The *backward diode* is similar to the tunnel diode in having heavily doped p and n regions, but does not have a negative resistance characteristic. Its usefulness lies in the fact that its fairly linear reverse characteristic makes it an efficient rectifier for low voltage signals.

What is a zener diode?

Semiconductor diodes have a breakdown voltage – the reverse breakdown voltage – at which the junction fails, allowing a sudden large increase in current for a very small voltage change. The voltage at which the breakdown occurs is known as the zener voltage (see Fig. 8). The zener diode is designed to operate at a point beyond the breakdown voltage. As substantial current variations will produce only very small voltage changes across it, the zener diode is useful in voltage regulating circuits, as a reference device and as a protective device in measuring instruments.

Fig. 43. A simple zener diode circuit. Note that the diode is connected with its cathode to the positive voltage end of the supply. The resistor R must be included so as to limit the amount of current which can flow through the diode

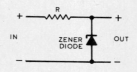

Fig. 43 shows a typical zener diode circuit. There must always be a resistor present to limit the amount of current which can flow through the diode, and in addition, the voltage which is used to operate the diode (the unstabilised or unregulated voltage) must be several volts higher than the zener voltage. A simple zener regulator of this type can be used to supply single stages which must use a stabilised voltage. When greater currents must be drawn, the zener diode can be used to control a power transistor. Fig. 44 shows a simplified example in which the zener diode controls the base voltage of a power transistor whose emitter voltage will then follow the base voltage (with the usual 0.6 V or so difference in d.c. levels). More elaborate circuits are used in stabilised power supplies and these also make use of voltage reference diodes, which are high-performance zener

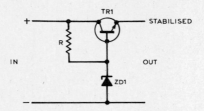

Fig. 44. A zener diode used to control the base voltage of a power transistor. This arrangement can pass much more current than the simple zener diode circuit of Fig. 43

diodes, set usually to 5.6 V. The reason for using this voltage is that a zener diode constructed to stabilise at this voltage is temperature-stable – the output stabilised voltage hardly changes as the temperature changes.

What is a varicap diode?

A reverse biased pn junction has a certain capacitance since it consists of two conductors, the p and n regions, separated by an effective layer of insulation, the depletion region, between them. This property is exploited in the varicap or varactor diode which is made with a large junction area. The capacitance varies with variation in the reverse bias applied, this feature being useful in a.f.c. and automatic (i.e. voltage controlled) tuning applications.

Varicap diodes are very extensively used for television and f.m. radio tuning. Many more tuned circuits can be 'ganged' (so that they can be altered in step) when varicap diodes are used for tuning, because the tuning is carried out by altering the bias voltage on the diode. In addition, because no mechanical movements are needed during tuning, much more compact tuners can be constructed with less risk of instability. Because no mechanical connections are needed between the tuner circuits and the tuning knobs (which control d.c. potentiometers), the tuner can be located anywhere in the receiver.

What is a unijunction transistor?

The unijunction transistor, or double-base diode, is a three-terminal device exhibiting a stable negative resistance region under certain conditions. It consists of an n-type silicon bar which has two base contacts, one at each end, and a p-type emitter region with contact at one side. Current flow between the base contacts is controlled by a current fed into the emitter.

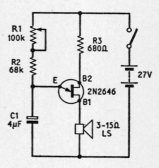

Fig. 45. Simple unijunction transistor metronome oscillator giving 90–220 beats per minute. To reduce the frequency of oscillation increase the values of R1, R2 and C1; to increase the frequency reduce the values of R1, R2 and C1.

C1 charges via R1 and R2 when the transistor is cut off. When the charge reaches a certain value the transistor switches on, C1 discharging rapidly through it. A pulse output is thus available at the base contacts and a sawtooth output at the emitter

The name of the device arises from the fact that it is a transistor with a single p-n junction. Forward biasing the junction provides a negative resistance characteristic between the emitter and one of the base contacts. The device is useful as a simple oscillator (see Fig. 45).

What is a thyristor?

As shown in Fig. 46 this device has four layers of semiconductor material, forming three p-n junctions. It differs from a diode mainly because of an extra connection, known as a *gate* electrode.

The thyristor presents a high resistance to current flow in either direction, but the resistance falls suddenly to a low value if the applied voltage in the forward direction exceeds a certain value. Then the action is that of a rectifier, with current flow in the forward direction only.

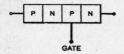

Fig. 46. The thyristor has four layers of semiconductor material and a gate connection

The gate introduces another factor. A small voltage applied to the gate switches the controlled rectifier to its low-resistance condition: that is, the gate pulse 'triggers' the device, so that it then acts as a rectifier. The controlled rectifier remains in this state after the pulse has been applied, provided that the forward current is not interrupted. If the current is turned off, however, the device returns to the high-resistance condition until triggered once more.

Thus control is possible either by varying the applied voltage in the forward direction, by varying the voltage applied to the gate, or by a combination of each.

The thyristor is used to a very large extent for controlling the average power delivered to a.c. loads, so that it finds many applications in light dimmers and in motor speed control. In addition to these uses, the thyristor is also used as the solid-state equivalent of a relay.

What is a triac?

Triac (a registered trademark) is the name given to a device which is virtually two back-to-back thyristors. A triac differs from a single thyristor in being able to conduct in either direction when the gate is triggered, so that triacs have now replaced thyristors in many applications which require the control of a.c. at comparatively low powers. Triacs have not, however, been made in the very high-power sizes.

What is a thermistor?

The thermistor is a device made from semiconducting materials with no junctions. Its value lies in its temperature sensitivity, because the resistance of a thermistor is very greatly affected by temperature. An n.t.c. (negative temperature coefficient) thermistor has a resistance value which is high at low temperatures but which drops to much lower values as the temperature increases. A p.t.c. (positive temperature coefficient) thermistor is the opposite type; its resistance is low at low temperatures and increases greatly as the temperature is raised.

Thermistors are used in temperature controllers, and in temperature measuring equipment, and also for the amplitude control of low-frequency oscillators. In this last application, the oscillator output is fed to a thermistor. If the current is sufficient to heat the thermistor, the change of temperature will cause a change of resistance. When the n.t.c. type of thermistor is used, this will in turn cause the resistance of the thermistor to drop, and the current to increase, and this current is used to bias the oscillator back, restricting the amplitude. In this way, the amplitude can be controlled to a stable level.

What is a 'trigger' diode?

The trigger diode is a two-terminal p-n-p-n device with a controlled breakover voltage. With reverse voltage it has the characteristics of a normal diode; in the forward direction it is non-conducting until a critical voltage is reached. At this point it switches, through a negative resistance region, to a conducting state, then behaving like an ordinary diode. Thus the device has uses in industrial switching circuits. Power versions are available for industrial a.c. power control equipment.

Trigger diodes are extensively used with thyristors and triacs to ensure that the triggering of these devices is sudden rather than gradual.

How does a phototransistor differ from ordinary transistors?

The phototransistor is essentially a junction transistor with a light-sensitive base. It is used as an amplifier and can therefore

63

be made to operate, via a relay, various kinds of equipment. The possibilities are numerous and quite widely exploited. For instance, it can be associated with burglar alarms, door-opening equipment, machinery controls and so on.

Warning devices can incorporate a phototransistor and infra-red source, the latter taking the form of an under-run car headlamp bulb fitted with an infra-red filter. The beam is then invisible. Applications of phototransistors in industry and in motor car accessories are mentioned later.

What material is used in solar cells?

This type of cell usually employs silicon crystals, and the arrangement is such that energy can be released due to a photo-voltaic effect. The junction is of p-type and n-type silicon, and one layer is made thin enough to permit solar radiation to pass through and irradiate the junction. The result is that charge carriers drift across the junction, thereby releasing energy. The sun's radiation is a mixture of infra-red and ultra-violet light, as well as visible light.

A lot of material is required, and therefore a solar 'battery' of cells is heavy and costly; other sources of energy are consequently under investigation. Immediate inquiries are however still concerned with semiconductors: as in some other devices, the use of evaporated films, rather than crystals, is likely to provide the most satisfactory result in relation to cost, weight and manufacturing problems.

How can semiconductors produce light?

When an electron and a hole meet, they can recombine, releasing energy equal to the amount of energy which was originally used to separate them. This energy is generally released in the form of heat, but when the amount of energy falls within the correct range of values and the semiconductor material is transparent, the energy can be released as visible light or as invisible infra-red or ultra-violet radiation. Crystals of materials such as gallium arsenide or phosphide and indium

arsenide or phosphide are used in the manufacture of l.e.d.s –
light emitting diodes, and for the l.e.d. segment displays which
are used in some calculators. Another form of display uses liquid
crystal materials which, though they do not emit light, can be
made either to transmit light or reflect light according to the

*Fig. 47. Using an l.e.d. A current-limiting resistor R must be
connected in series otherwise the l.e.d. will be damaged*

Table 2

LED	LCD
Bright display, visible in dim light	Reflective display, visible in bright light
Needs high current, several mA per segment	Very low current required, few μA for each whole display
Easy connection to transistor or i.c. circuits	Needs a.c. high frequency signals which are generated by an i.c.
Long life	Life uncertain, can be adversely affected by direct sunlight or high temperatures
Inexpensive	Comparatively expensive

strength of the electric field across the material. Table 2 shows some of the comparative advantages and disadvantages of l.e.d. and l.c.d. displays.

What is an integrated circuit?

A circuit in which all components and interconnections are formed within a single piece of semiconductor material. This is largely a development of silicon planar technology: in addition to diffusion to produce the various sections of transistors, in an integrated circuit other parts of the circuit, e.g. diodes and resistors, are produced by multiple diffusion. Further additions may be made by surface deposition on the wafer. In this way complete circuits, e.g. multivibrators, small amplifiers, etc. can be produced as a single entity, the technique enabling large quantities of the circuits to be simultaneously produced in microminiature form. Since active components such as diodes and transistors take up very much less space in an integrated circuit than passive components such as resistors and capacitors, much greater use is made of active components in integrated circuits than in conventional circuits where separate components are assembled together.

What is meant by LSI?

LSI means large-scale integration, in which a circuit which would normally require thousands of transistors and other components is made in the form of a single i.c. LSI circuits use field-effect transistors in i.c. form, and are the devices which have made pocket calculators and desk-top computers possible.

6

TRANSISTORS IN INDUSTRY

What are the more important uses of semiconductor devices in industry?

These devices are of course used in thousands in computers engaged on a variety of tasks in industry, commerce and research. Again, transistors offer advantages in measuring instruments and in control systems for electric motors, generators and manufacturing processes. Where large numbers of circuits are involved, small size and moderate demands on power supplies are most desirable attributes. Among the many types of device used, thyristors, light-sensitive transistors and silicon rectifiers and transistors are especially important.

What basic circuits are used?

In addition to basic oscillating and amplifying arrangements, noted earlier, transistor circuits for switching and counting are among the most important in industrial equipment and computers. An electronic switch can be operated very quickly and has no moving parts. A transistor that is 'on', i.e. conducting, is equivalent to a closed switch, whilst one that is 'off', i.e. non-conducting, is equivalent to an open switch. The following examples will serve as an introduction to the subject.

Although valves were formerly used for electronic switching, or gating as it is often called, the transistor with its faster speed of operation is more suitable for this task and has in fact made possible the development of the advanced types of equipment (notably computers) which are becoming familiar today.

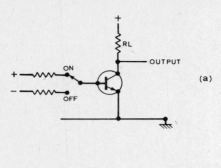

(a)

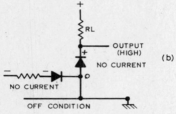

(b)

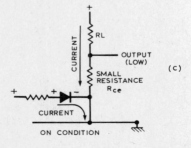

(c)

Fig. 48. Equivalent circuits for common-emitter transistor used as a switch (a), showing the off (b) and on (c) equivalents

68

In pulse circuits the transistor is employed as a voltage or current switch. The timing of its operation depends on the values of other components (resistors and capacitors, for instance), and the pulse repetition frequency is determined by the speed at which the transistor will switch from one condition to another.

For convenience the two junctions of the transistor can be regarded as separate diode rectifiers. A common-emitter switch circuit is shown in Fig. 48(a), and the equivalent diode circuits for the transistor's off and on conditions are shown in Fig. 48(b) and (c). In the off condition both diodes are reverse biased and will not conduct. No current flows through the load R_L, so that the output voltage is identical to the supply voltage. When the base is forward biased, the base circuit conducts, and the equivalent circuit between the collector and the emitter is a small-value resistor. The potential divider formed by the load resistor and the resistance between the collector and emitter, R_{ce}, causes the output voltage level to drop to a very low value, typically 0.2 V. Note that this is less than the value of voltage between the base and the emitter.

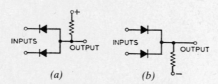

Fig. 49. Two-input diode AND and OR gate circuits

Again using diodes in a simple example it is possible to consider the action of AND and OR gates, which are among the essential elements of computer logic circuits. Referring to Fig. 49, the AND gate gives an output only when signals are applied simultaneously to each input. The OR gate gives an output if a signal is applied to either one or both of the inputs. When the input signals are positive, (a) acts as an AND gate and (b) as an OR gate; the converse applies when the signals are negative. In practical cases there would be several inputs – not just two as

69

shown here. Another common arrangement is a NOT gate, which is an inverting amplifier.

Logic circuits in the most recent computers may be relatively elaborate – though physically small – and make use of different types of component; they operate at speeds measured in nanoseconds (= thousandths of a millionth of a second).

These modern logic circuits make use of transistors in integrated form; the use of separately wired transistors and diodes has now died out except for some prototype work.

What other pulse circuits are used?

A very common circuit is one which switches between the two operating states, on and off. Either or both states may be *unstable* – that is the circuit will not remain in the state called 'unstable' but reverts automatically to the other state. If neither

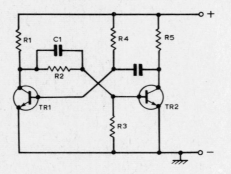

Fig. 50. Basic monostable circuit. In the stable state, TR1 conducts, and the low voltage at the collector of TR1 prevents TR2 from conducting. A positive pulse at the base of TR2 will cause this transistor to conduct, shutting off TR1 until C2 can charge through R4

70

state is stable the circuit is referred to as being *free running*, i.e. it continually changes from one condition to the other.

The multivibrator was first devised to meet the need for a generator of square waves, which are rich in harmonics. The basic arrangement shown in Fig. 28 has two unstable states. When the circuit is brought into operation, slight unbalance in the components causes the transistors to go into one of the unstable states: subsequently there is oscillation between the two unstable states.

The monostable circuit (Fig. 50) has one stable and one unstable state. The name monostable is used because the circuit

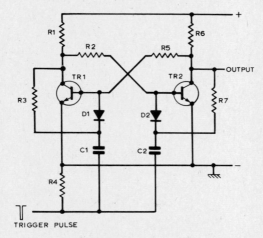

Fig. 51. The bistable, with steering diodes. Each steering diode has its cathode biased from the collector of a transistor and its anode biased from the base of the same transistor. A transistor which is switched on will have its collector voltage at 0.2 V and its base voltage at 0.6 V, so that its steering diode is almost conducting. By contrast, the other diode will be considerably reverse-biased, so that only the conducting transistor is affected by the trigger pulse

71

has only one stable state (with TR1 conducting) to which it returns after being briefly switched over by a trigger pulse. The time for which the monostable is in its unstable state is determined by the time constant of R4, C2 in the example of Fig. 50.

The Eccles-Jordan bistable arrangement (Fig. 51) is especially important in view of its wide use for counting purposes. One input pulse throws the circuit into one stable state, where it remains until a second pulse returns it to the first condition. For every two input pulses there is one output pulse: thus the circuit divides by two.

In the example shown the bistable circuit is triggered on and off with pulses of the same polarity. Switching would not occur if the pulses were applied at the same instant to both transistors, and therefore the 'steering' diodes D1, D2 are incorporated to avoid this. The arrangement is such that a negative-going input pulse is always steered to the 'on' transistor (though reversal of diode connections makes the circuit sensitive to positive-going pulses instead).

In practice a series of these circuits will be assembled so that the output from one operates the next. In this way a binary counter (or, with modifications, a decade counter) can be formed. Information is carried out of a bistable circuit via a gate, for example the AND gate already mentioned.

The bistable circuit is commonly referred to as a 'flip-flop' in practice, and is invariably obtained in integrated form.

What is a d.c. amplifier?

Amplification of a d.c. input is often required in, for instance, measuring instruments, computers and automatic control systems. The normal a.c. amplifier, incorporating inter-stage capacitor (or transformer) coupling, is not suitable for this: the capacitor, having some specific time constant, will block the signal. Therefore the stages must be direct coupled.

In the d.c. amplifiers which are used for meters, there are problems of maintaining correct bias, of avoiding variations in gain, and in the drift of the zero reading which causes the meter

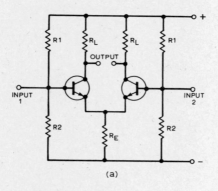

(a)

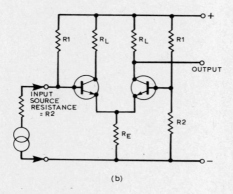

(b)

Fig. 52. Basic balanced amplifier (differential amplifier) circuits. (a) Push-pull, or fully-balanced version, (b) version with single-ended input and output

73

to show a reading when there is no input. The first two can be tackled by using a suitable circuit, a balanced amplifier (Fig. 52) with negative feedback of d.c. to stabilise gain and bias. The problem of drift has to be tackled by careful design and selection of components.

Circuit designers are now helped by the availability of double transistors – that is, a pair of transistors in one encapsulation. This provides the important advantage of a minimised temperature difference between the pair and in general supports the development of circuits with reduced drift.

One important use of d.c. amplifiers is as 'operational amplifiers' in analogue computers. As the name suggests, the operational amplifier is one used to perform mathematical operations such as summation, integration and differentiation. This it does by means of feedback. The basic arrangement consists of an amplifier having a series input impedance Z_1 and a parallel feedback impedance Z_2. Provided the gain is sufficiently high, the input and output voltages e_1 and e_0 are related as follows: $e_0/e_1 = -Z_2/Z_1$. By varying the series and feedback impedances, which may be resistive, reactive or non-linear, different operational functions can be achieved. The fact that such an amplifier requires high gain, stability under various feedback conditions, high input impedance and low drift means that it is also attractive for instrumentation and other such applications.

Operational amplifiers in integrated form, such as the well-known 741, are now preferred to discrete (separate transistors) circuits.

What is 'environment stabilisation'?

In the interests of complete reliability (or as near to perfection as it is possible to get), equipment incorporating transistors and other temperature-sensitive devices can be placed in an enclosure and provided with closely controlled working conditions. For instance, transistorised assemblies for aircraft may be protected in this way. All the enclosed components are kept at a temperature that is controlled to within a fraction of a degree, humidity is controlled, and there is a high standard of thermal insulation. A refrigeration unit may be included.

Thus the stabilisation of the equipment's environment is achieved. The provision of extra materials adds to the initial cost, and it might be thought that added bulk would be a disadvantage. But in fact such close control of operation leads to reduced power consumption, some simplification of circuits and further opportunities to reduce physical size. Moreover greatly improved reliability leads to welcome economies in activities, such as aircraft operation, which are already costly. It can therefore be expected that such techniques will be used more often in the future.

A less advanced method of protecting transistorised equipment is to arrange for simple air conditioning. For example, aircraft navigational aids have been built as sealed modules filled with inert gas slightly above atmospheric pressure to provide protection against dust and impurities in the air. The methods mentioned here can be applied to computers.

What types of rectifier are used in electrical engineering?

Here again, silicon semiconductor devices are widely used. Many industrial applications formerly used metal rectifiers which employed selenium or copper oxide cells, which were more resistant to damage than germanium diodes but which have now been displaced by silicon diodes. Silicon junction rectifiers have the desirable characteristics of low forward resistance, very high reverse resistance, and the ability to function at high temperatures, as well as being smaller and more robust than the older types of rectifier.

For voltage control, as in the speed control of d.c. and a.c. motors, thyristors are now widely used. Other applications for semiconductor devices, particularly of the silicon variety, are in the control systems associated with machine tools and alternators.

The thyristor has been described in Chapter 5. Compared with the thyratron, which it is tending to replace in industrial plant, the thyristor has a very low forward voltage drop in the conducting state and is therefore highly efficient. It has trigger and recovery times which are measured in microseconds, and it can work at a high temperature. And, of course, compared with a valve device, it has no heater – and therefore no warming-up period – and is more compact and robust.

A gate pulse triggers the device to conduct in a forward direction: it is 'blocked' until this happens. It remains conducting until the current in it is turned off. The device can be fired at any point in the a.c. half-cycle, so enabling a continuously variable d.c. output to be obtained from a constant a.c. input.

What is a typical thyristor application?

Infinitely variable-speed drives are required for many kinds of plant, including printing and textile machines, pumps, large fans and machine tools. A fairly simple yet flexible control system can be designed around thyristors.

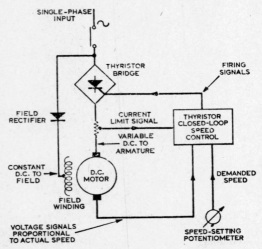

Fig. 53. Variable speed drive for a d.c. motor (block diagram)

The general arrangement of a system for a small d.c. motor is shown in Fig. 53. The motor field is supplied at constant voltage by rectified a.c. The speed of the motor is altered by varying the armature voltage using a thyristor bridge circuit, which is switched from a non-conducting to a conducting state by low-power firing pulses. The timing of the pulses controls the instant during the a.c. cycle when the thyristor starts to conduct and thus controls the average output voltage level.

A simple control circuit adjusts the timing of the firing pulses according to that demanded by the speed-setting potentiometer. It also compensates for load increases and prevents overloading.

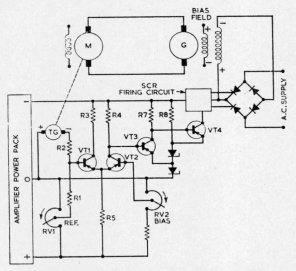

Fig. 54. Simplified circuit of a transistorised Ward Leonard control system

A typical high-power application of thyristors is shown in simplified form in Fig. 50. This is a Ward Leonard motor-generator set, in which the generator field is supplied from thyristors. The arrangement allows for regenerative braking when the set is adjusted to a lower speed. The thyristor firing circuit incorporates a saturable reactor requiring a current output from a transistorised control amplifier (VT1-VT4 in the circuit).

Why are heatsinks used in industrial equipment?

Reference has already been made to the importance of conducting heat away from the transistors used in even small items of equipment (such as audio amplifiers). In industrial plant where large silicon rectifiers and thyristors are used, very substantial heatsinks with large surface areas are required. These may be made from extruded or diecast aluminium or other materials.

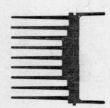

Fig. 55. Cross-section of a heatsink for a silicon rectifier

In using heatsinks designers of equipment have to take into account many factors such as component tolerances, the possible accumulation of dust on the heatsink when it is in use, the ambient temperature and restricted ventilation. In some cases forced air cooling is employed. A typical extruded heatsink cross-section is shown in Fig. 55. This illustrates the large radiating area provided by fins.

How are light-sensitive devices used in industry?

The photocell is associated with a projector – a lamp with some means of directing the light beam – and the two components are placed so that an object to be detected will pass between them.

78

The cell may be a phototransistor or some other photoconductive device (for example a cadmium sulphide cell) with separate transistor amplifier. Whatever the arrangement, the circuit normally includes a relay to operate other equipment (see Fig. 58). It can be arranged for this unit to function either when the light beam is interrupted or when it is applied.

This simple equipment has a remarkably wide variety of uses in industry. These include the detecting, counting and monitoring of articles passing on a production line; the counting of vehicles or people; and the initiation of such actions as the opening of doors, the operation of alarms and the control of machines. Photocells can be used for timing purposes – in car trials, for instance.

Operation can be made to depend on gradual changes of light value (rather than on/off conditions), so that it is possible to detect smoke or cloudy substances. An interesting application is the monitoring of turbidity and consistency of fluids. Again, the equipment can take the form of an automatic controller for lighting systems, operating when natural light falls to a certain level.

It should be noted also that specialised types of device have an important place in the steel industry. Hot steel is a source of infra-red radiation, and therefore these devices can be incorporated in control and safety systems.

There has been increasing interest in photo-detector devices for the detection of radiation in the region between infra-red and microwaves (wavelengths of about 100 micrometres and above). This type of device is required for research and in radiocommunications. Some detectors for the near infra-red have employed materials which are not doped with impurities: indium antimonide, lead sulphide and tellurium have been used. One practical detector for research purposes has incorporated n-type indium antimonide.

To what extent are semiconductor devices used in instruments?

Transistors are widely used for such purposes as oscillation, amplification, the operation of control circuits (for instance in automation schemes) and the generation of signals, and have to a great extent supplanted valves in a variety of voltmeters, frequency meters and other essential measuring instruments. Practical advantages include the reduction of size of instruments, the virtual elimination of ventilation problems and decreased demands on built-in power supplies. Printed circuits, and in some instances the use of microcircuits, should also be mentioned.

For measurements at very high frequencies special types of transistor offer distinct advantages over valves: the higher value of mutual conductance permits the use of a lower load resistance, and a wider operating bandwidth is also possible. On the other hand, high-power transistor a.c. amplifiers with high input impedances are not easily devised. Recent types of special-purpose transistor have improved the situation.

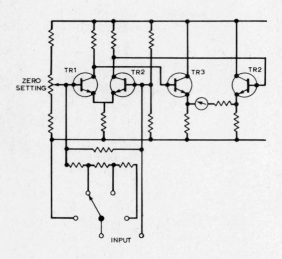

Fig. 56. Basic circuit of a transistor voltmeter

In the circuit of a transistor voltmeter shown in Fig. 56, the first two transistors form a balanced voltage amplifier. These feed a pair of transistors which, in the emitter-follower connection, provide additional current amplification. High-value resistors are switched into circuit to provide high voltage ranges on what is basically a very sensitive meter. To minimise temperature effects in such an instrument balanced pairs of silicon transistors are used.

Also important in instruments are pulse-generating and counting circuits and d.c. amplifiers, to which reference has already been made. Again, in general the transistor scores over the valve.

Perhaps the most unusual and interesting phenomenon that can be exploited in measuring instruments is the Hall effect, which occurs when a piece of semiconductor material is held in a magnetic field. Briefly, if current is passed through a wafer of semiconductor and the magnetic field is perpendicular to the wafer (passing between its faces), a voltage is produced between contacts on opposite edges of the wafer. The voltage is proportional to the product of the magnetic field and the current through the wafer.

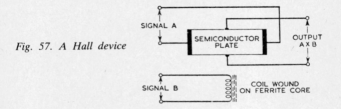

Fig. 57. A Hall device

In practice a Hall device consists of the wafer mounted in the gap of a ferrite core on which a coil is wound. These two elements can be represented as shown in Fig. 57. Input A passing through the semiconductor interacts with the magnetic field produced by input B. The output voltage, from the edges of the

81

semiconductor wafer, is proportional to the product of the two input signals.

The Hall device can be used to convert the measurement of magnetic flux into a measurement of either d.c. or a.c. voltage. It can therefore be regarded as a transducer. In another application, if one input is a.c. and the other d.c., the output is an alternating voltage of proportional strength to the d.c. input. In other words there is d.c. to a.c. conversion. Hall devices can be used in computers and radio equipment as well as in measuring instruments.

How does a transistor increase relay sensitivity?

The relay itself enables a small current to switch circuits carrying much larger currents. A relay actuated by a current of, say, 10 mA may control circuits carrying 10A. However, a considerable

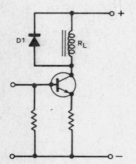

Fig. 58. Use of a transistor to increase relay sensitivity

increase of sensitivity can be obtained by using a transistor as a current amplifier, so that the input current need be only a few microamps.

As shown in Fig. 58 the transistor base is returned to the emitter through the resistor, and the collector current is insufficient to operate the relay, R_L. With a small input current the transistor conducts and the relay operates. It should be noted that the transistor is using the supply already provided for the relay, and therefore the arrangement is a convenient one. It is widely used in industrial equipment.

The diode which is shown connected across the relay coil must not be omitted. When the relay is switched off, the inductance of the relay coil will cause a large surge of voltage to be generated which can destroy the transistor. The diode shorts out this voltage to the supply line, thus preventing damage.

Is there a simple instrument for the measurement of radioactivity?

A Geiger-Muller tube and ratemeter, enabling a reading of radiation intensity to be obtained, can be made in a fairly simple form using transistors. The requirement is to average the count from a Geiger-Muller tube and to amplify and convert this into a reading on a meter. A suitable circuit is shown in Fig. 59.

The circuit comprises a type HC4 tube; a two-stage amplifier with an output suitable for a loudspeaker or earphone as well as for the meter; an ordinary micro-ammeter provided with switched ranges; and an integrator to average the count rate. The power supply section provides the low voltage for the transistors as well as the high voltage for the Geiger-Muller tube. Battery-operated designs use an inverter, a transistor oscillator, step-up transformer, and rectifier arrangement, to supply the 400V d.c. for the G-M tube.

Can electricity be produced from semiconductors on a large scale?

The prospect of generating electricity without the intervention of machinery such as turbines and alternators is of course an attractive one, and the various methods tried on an experimental scale have included the use of semiconductors. Certain devices, such as solar cells, are of course beyond the experimental stage, but the amount of electricity obtained from a typical battery is small.

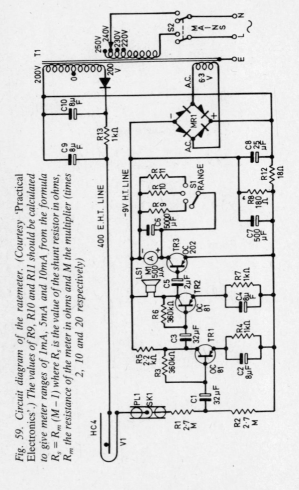

Fig. 59. Circuit diagram of the ratemeter. (Courtesy 'Practical Electronics'.) The values of R9, R10 and R11 should be calculated to give meter ranges of 1mA, 5mA and 10mA from the formula $R_s = R_m (M-1)$ where R_s is the value of the shunt resistor in ohms, R_m the resistance of the meter in ohms and M the multiplier (times 2, 10 and 20 respectively)

One approach which appears to hold promise is the conversion into electricity of heat from nuclear reaction. Russian scientists have reported success with an experimental reactor, in which the heat generated is conveyed to a semiconducting thermo-electric converter. The converter, containing elements of silicon-germanium alloy, is located on the external surface of the reactor. One side of the elements receives heat while the other is cooled, and thus an electric current is generated. It remains to be seen whether large-scale generation will be economically and technically feasible.

7

OTHER USES FOR
SEMICONDUCTOR DEVICES

What other important uses are there for transistors?

Important fields of application include motor vehicles, space research, and medicine. The trend to automatic operation in cinema projection involves the use of semiconductor devices. Electronic telephone exchanges will depend increasingly on semiconductor devices of various kinds; so will repeater amplifiers associated with telephone cables, such as those laid under the sea, which must have a long, trouble-free life.

To what extent can semiconductor devices be used in vehicles?

Apart from the familiar and obvious applications in car radio, there are many existing and possible uses in the vehicles themselves. So far it is the heavier vehicles that have benefited the most, but it is likely that considerable use will be made of transistors and other devices in cars. Such developments, in road transport as a whole, are obviously of great economic and technical importance in view of the enormous number of electrical units that are required. It is therefore worth reviewing them in some detail.

In vehicles semiconductor devices are used in the interests of improved performance, consistency and reliability of electrical systems. The temperatures which are encountered, continuously or occasionally, are so high that careful heatsinking of silicon transistors, diodes and i.c.s is needed, and also some attention to the airflow over the heatsinks.

Important advances have been made in electrical generators, ignition systems and associated control equipment, and there are various items of auxiliary equipment in which semiconductors are likely to be used more frequently.

In most of these applications, integrated circuits have replaced discrete transistors because of their compact size, high reliability and low cost. The principles which are involved, however, are the same irrespective of how the devices are constructed, though the use of i.c.s enables us to apply circuits which would be much too complex to construct using discrete transistors.

How has the vehicle's generator developed?

Although the a.c. generator is a familiar item in motor cycles, d.c. generators (dynamos) have long been used in cars and many heavy vehicles. They use brushes which need maintenance and eventual replacement, and there is a limited range of engine speeds over which the batteries can be charged. A disadvantage known to car-owners is the absence of battery-charging when the engine is idling – a state of affairs which becomes more common as traffic density increases. Modern conditions evidently mean that older types of electrical system are no longer entirely adequate.

A great improvement can be made by using instead an a.c. generator (alternator) with a rectifier. An alternator needs no maintenance apart from a very occasional routine inspection; it can be smaller; and semiconductor devices can now be used in its output circuits (instead of the relatively heavy metal rectifiers that would otherwise be needed). One development is the three-phase alternator with which silicon diodes are used for rectification: in fact it is possible to incorporate rectifying diodes in the alternator itself. Such equipment is now appearing in cars, bringing the benefits of light weight and charging at idling speeds.

Associated with the use of the alternator and silicon diodes is the modern control circuit which replaces the old electromechanical device. A typical example is shown in Fig. 60. The operating principles are as follows.

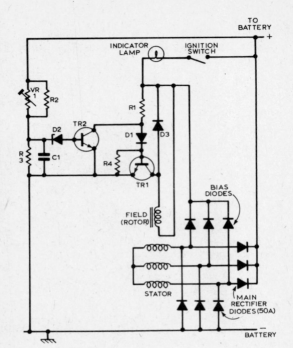

Fig. 60. A modern alternator circuit (courtesy of AC Delco). The entire regulator apart from the main rectifier diodes and bias diodes is constructed as a thick-film circuit on a ceramic block

When the ignition switch is closed, current flows through the indicator lamp, and biases on TR1, so that current can also flow through the field winding of the alternator. Whenever the engine turns, the stator windings will generate a voltage which is rectified by the diodes. The main rectifier bridge is used to feed the charging current to the battery, but an independent set of three diodes is used to supply a bias voltage to the point where

88

the resistor R1 is connected to the indicator lamp. As this voltage rises, the indicator lamp dims and eventually ceases to glow.

The base of TR2 is biased from the battery terminal through the large-values resistor chain, R2, R3, but the zener diode prevents current from flowing into the base of TR2 when the generator is not charging. As the generator voltage increases, however, the zener conducts, so that TR2 conducts, and the bias to TR1 is shorted out. This reduces the current in the field coils, so that the generator voltage drops. When the voltage drops low enough, the zener diode ceases to conduct, and the cycle repeats. This on/off form of control avoids the problems of high dissipation which would occur if the transistors were operated in class A. When the battery needs a large current, the circuit spends most of its time with TR1 fully on; when the battery is nearly fully charged, TR1 is off for most of the cycle.

Are there other uses for the zener diode?

A zener diode can be used in a simple regulating device for a motor cycle's 12V a.c. electrical system. The diode is connected across the battery and acts as a by-pass, passing rectified current from the alternator according to the battery's state of charge. As the battery becomes recharged its voltage rises; when it reaches about 14V the diode, so far in a non-conducting (high resistance) condition, becomes partially conductive and provides an alternative route for part of the alternator output.

Further small rises in battery voltage lead to large increases in diode conductivity until, at about 15V (the on-charge voltage of a fully charged battery) most of the alternator's output is by-passed and the off-load voltage is stabilised.

If now a headlamp is switched on, the system's voltage falls below 15V; less current flows through the diode and the balance is diverted to supply the load.

If a particularly heavy load takes the voltage below 14V, the zener diode reverts to its non-conducting state and all the output of the alternator is used to meet the demands of the battery and equipment.

How is the transistor used in a car ignition system?

The nature of the conventional ignition system for cars is well known: important components are an induction coil and a contact breaker which is housed inside the distributor. The contacts have to break a fairly large current at a rate of up to several hundred times per second. Sparking and mechanical wear mean that periodic adjustment – part of the general tuning of the engine – is essential if engine performance is to be maintained.

Most modern transistor ignition circuits use capacitive-discharge methods which have replaced earlier circuits. An inverter circuit generates d.c. at about 400V, and this is used to charge a capacitor through a resistor. At the time of ignition, a thyristor is triggered, allowing the capacitor to discharge into the primary winding of the ignition coil (using a conventional 12V coil) and so causing a high voltage to be induced at the secondary. This is the reverse of the conventional system in which breaking a current in the primary causes the induction of the high secondary voltage.

This system has the advantage that it can be used with or without conventional contact-breakers. When conventional 'points' are used, the opening of the points allows the base of a transistor to be biased on, and the amplified current is used to trigger the thyristor. Though the electrical sparking at the points is almost completely eliminated by this method, mechanical wear is not, and a more refined system uses a revolving vane which interrupts an infra-red beam from an l.e.d. to a phototransistor. The phototransistor output is amplified and used to trigger the thyristor.

It is important for a car ignition system to be constructed from very high quality components, since a sudden failure can have fatal results. Such circuits should be bought only from reputable manufacturers, and any kits which contain unmarked components should be avoided. A common source of trouble is breakdown of the capacitor, or failure of the inverter. If this occurs while the vehicle is overtaking the results can be catastrophic.

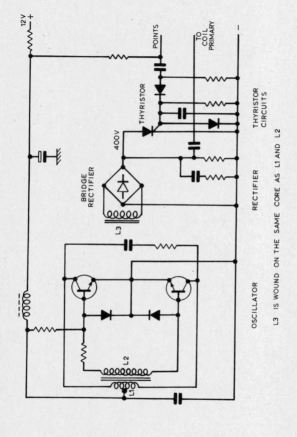

Fig. 61. A modern capacitor-discharge ignition circuit

91

One considerable advantage of electronic ignition is that the timing of the spark can be controlled completely electronically, so that microprocessor control of ignition becomes possible, and is being incorporated into cars in the USA. It is important to note that these are systems whose reliability, unlike that of add-on or home-brew systems, has been checked by millions of miles of test driving under all possible conditions.

What other uses are there for transistors in cars?

Possible uses are in automatic transmissions, the control of air conditioning systems and, eventually, control equipment associated with petrol-injection systems (which are very little used at present). There are many opportunities for using transistors and diodes in test equipment for garages. The transistorised clock is an obvious application, although the trend seems to be towards digital clocks driven by i.c.s.

How can light-sensitive devices be used in cars?

One example is an automatic anti-dazzle rear-view mirror for night time driving. Like some manually adjusted mirrors it is of the prismatic type but in this model a photocell, which detects glare from following vehicles, is connected to a printed wiring assembly incorporating three transistors. The amplified current actuates a solenoid which in turn moves the mirror.

Another device uses a photocell to control the switching of side or parking lights according to the natural light conditions. There are several commercial models. A possible further application, technically difficult and not yet exploited, is in automatic headlamp dipping.

How can a high-voltage supply be derived from car batteries?

A high-tension supply can easily be arranged using a transistorised inverter. One possible circuit for a light-duty converter is shown in Fig. 62. It is designed to give a d.c. output of 300V at up to 130 mA from an input of 12V d.c. The output of the 12V

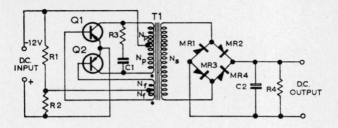

Fig. 62. 12V to 300V transistor d.c./d.c. converter (Courtesy Newmarket Transistors Ltd)

Q1 NKT402	R1 25Ω	Resistors 10% tolerance
Q2 NKT402	R2 1Ω	MR1-4 S.T.C. type
C1 4μF	R3 7·5Ω	280LU-679A or
C2 32μF(350 V)	R4 22k(2W)	equivalent, 300 V 150 mA.

push-pull square-wave oscillator is transformer-coupled to give an a.c. square-wave voltage of 300V. This is rectified by the metal rectifiers MR1–MR4 and smoothed by the $32\mu F(C2)$ electrolytic capacitor.

Can semiconductor devices be used by the railways?

One use is in a system of automatic train control, in which signals sent to various parts of the line can be made to convey information to a passing train. The equipment makes use of the semiconductor indium antimonide. This substance exhibits a Hall effect: if a magnetic field is brought near it while a current is being passed, a voltage drop is developed across it. Another experimental application has been in signalling: the aim has been to replace electromechanical methods with computer-like techniques using plug-in transistorised units.

To what extent are semiconductors used in space research?

Some of the most advanced semiconductor devices are used in earth satellites. Space experiments for research purposes, such as those concerned with the study of the ionosphere, involve a wide variety of instruments, and techniques are now so well advanced that a remarkable amount of engineering can be packed into a small container. The equipment employs many hundreds of transistors, as well as arrays of solar cells to charge the batteries. A communications satellite is particularly complex with its equipment designed to receive, store, amplify and retransmit signals.

Miniature components and circuits are used in spectrometers, counters for measurement of particle size, telemetry equipment, tape recorders and many other items. The quality of the components and the construction generally must be such that reliable performance can be achieved in the face of difficult conditions – for example high vibration frequencies, wide variations of temperature, and accelerations of a kind that would not be encountered in other applications. Microminiature circuits, mentioned elsewhere, make a vital contribution.

Satellites are used in weather forecasting and photographs are taken at regular intervals showing the development of cloud patterns. The data are transmitted to earth-based stations and a good deal of international cooperation has been achieved. For instance the information from the north Atlantic satellites is exchanged and interpreted by the British and American weather forecasters and by both the air forces and navies of the two countries, supplemented by information gathered by ships and aircraft in mid-Atlantic.

The system is however not considered to be adequate and methods are continually being sought to make more accurate weather predictions which would be broadcast as a global service by the satellites themselves.

In this system, each satellite would incorporate a very small computer for data analysis and subsequent predictions. It is believed that the relevant branches of technology are now so advanced that they could produce the special types of micro-miniaturised circuits and semiconductor devices that would be required. An improved source of energy, possibly a nuclear-powered generator, would also be necessary.

Are transistors used in medicine?

Many of the possible uses are only now being explored, but already some notable advances have been made in medical electronics. A variety of instruments are available for both diagnosis and research. These are usually to aid study of the condition or behaviour of the body; for instance transistorised instruments are available to obstetricians for the observation of foetal heart waveforms. Instruments are used in operating theatres for the measurement of blood pressure, respiration, temperature and other fundamental matters. As usual, discrete transistor devices have now been replaced by i.c.s in most applications.

Photoelectric devices are important, for example, in biochemical analysis. A method of measuring heart rate uses a phototransistor and light source attached to an ear lobe.

The main contribution of semiconductor devices is in much needed miniaturisation of equipment. Measuring and telemetry instruments have to be carried on the person, and therefore small size and weight are essential. Probably the most spectacular example is the radio pill which contains, in a housing 2 cm long and 1 cm diameter, an oscillator, aerial, transistor, battery and other parts. The pill is swallowed and transmits data from inside the body about temperature, acidity, pressure and other factors. The frequency of radiation can be made to vary according to changing conditions inside the patient.

Another example of miniaturisation is the equipment used for certain cardiac disorders. A transistorised receiver can be inserted in the heart muscle of a patient whose heart beats irregularly. This electronic stimulator is controlled by a transistorised transmitter, worn externally. A high standard of reliability is important, since it may be necessary to leave the device in position for a number of years.

Physiological amplifiers, designed for electrical inspection of patients, gain in usefulness if they can be battery-powered and thereby made easily portable. Transistors have permitted a substantial advance in design. Again, a low-voltage circuit is safer. There is however the difficulty that such an amplifier

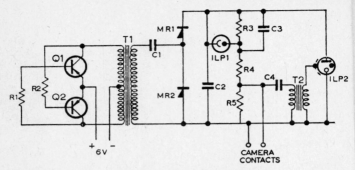

Fig. 63. 40 Joule, 6 V transistor photoflash (Courtesy Newmarket Transistors Ltd)

Q1	NKT404	C1	10μF(500 V)	R1	22Ω(1 W)
Q2	NKT404	C2	500μF(500 V)	R2	22Ω(1 W)
MR1⎱	500 V voltage	C3	0.05μF	R3	1.5M
MR2⎰	doubler	C4	0.25 μF	R4	3.3M
ILP1 90 V neon, ILP2 40 Joule				R5	3.3M
500 V max photoflash lamp.					

usually has a low input impedance, and the source impedance (the patient) under normal working conditions is high. A field effect transistor will provide the required input condition, and therefore equipments using a hybrid combination of transistors have been devised.

How are transistors used in a photoflash?

Transistors can provide the power for an electronic photoflash of professional type. The circuit shown in Fig. 63 is supplied by a 6 V battery and is designed to deliver 40 joules of energy at 500 V to a flash bulb, ILP2. The charge builds up to 500 V in about 40 seconds. When the photoflash is ready for use the neon lamp ILP1 lights up. The power supply consists of a self-oscillating push-pull converter with an output of 250 V which is raised to 500V (by the half-wave voltage doubling circuit MR1, MR2) for charging capacitor C2.

The unit is suitable for use with any camera provided with appropriate contacts. When the contacts are closed, the capacitor C4 discharges through transformer T2; this induces a large voltage in the secondary which fires the bulb. The energy stored in C2 is then liberated in the bulb to give intense illumination.

What types of battery are used in miniaturised equipment?

Special requirements, for example in connection with military equipment, often have to be met by individual designs, with characteristics adjusted to particular working conditions. However, in more familiar types of small transistorised product (hearing aids, for instance), silver-oxide cells now make an important contribution.

This type of primary cell has a low and uniform impedance, good low-temperature characteristics and a very linear discharge characteristic – that is, the voltage is constant with time. In some versions it may weigh only a small fraction of an ounce. It is well suited for powering silicon transistors.

Again, the cell's terminal voltage is higher than that of the mercury cell, another well-known miniature type. The open-circuit voltage is 1.6 V; with a typical current drain it remains as high as 1.5 V, compared with 1.3 V for mercury cells in similar conditions.

The silver-oxide cell consists of a depolarising silver-oxide cathode, a zinc anode of large surface area and a highly alkaline electrolyte. In hearing-aid cells the electrolyte is potassium hydroxide, which ensures maximum power density at low currents. In the extremely small cells used in watches, sodium hydroxide is chosen for long-term reliability.

Indium bismuth cells are also suitable for miniature equipment. Rechargeable nickel cadmium batteries are sometimes used in space satellites and can be charged from solar cells. This type of battery, often made in miniature form, can supply large currents and has a reputation for good performance at very low temperatures.

Portable television receivers make heavy demands on batteries. There is, however, a 12V battery which, with a special

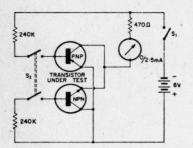

Fig. 64. A simple transistor tester for pnp or npn transistors, giving an indication of the gain provided by the transistor.

stabilising circuit, is capable of supplying 0.25 A for 40 hours when used for about two hours a day. Nickel cadmium rechargeable cells are also used in portable television sets.

Is there a simple transistor tester?

The arrangement shown in Fig. 64 may be used to make a rough check of transistor gain. With the transistor to be tested connected and S1 closed, the meter will indicate the collector leakage current – only a small reading should be obtained. Pressing spring-loaded switch S2 feeds a 25 μA current to the transistor base: by calibrating the meter with a full scale gain of 100, the transistor gain can be read off directly.

8

SERVICING TRANSISTORISED EQUIPMENT

What are the most common faults in transistorised equipment?

Transistors are robust and, if operated under correct conditions, are seldom themselves the cause of faults in transistorised equipment, though they may of course fail due to the breakdown of an associated component. Far more common than transistor failures are mechanical faults such as unsatisfactory switch contacts, 'tired' spring contacts in jack sockets, noisy volume controls, components being knocked off printed boards, broken aerial leads and cracked ferrite-rod aerials. The most common electronic faults are: battery failure; resistors going high; capacitors going leaky. The first check in battery-powered equipment should always be of the battery.

Is there a simple method of checking the battery?

A helpful method is first to measure the voltage across its terminals (V1), then to place a 100-ohm resistor across its terminals and measure the voltage again (V2). The current (I) in mA drawn by the battery is $10 \times V2$, and the battery internal resistance can be calculated from the following formula:

$$\frac{V1 - V2}{I} \times 1000 = \text{battery internal resistance (ohms)}.$$

A good 9 V battery will have an internal resistance of about 10 ohms. Up to 50 ohms the battery is workable, above this it should be replaced.

How should one proceed after checking the battery?

It is recommended that you first look carefully for mechanical faults of dry joints. If everything is in order, isolate the faulty stage by systematically testing backwards through the equipment, using a low-voltage signal injector (preferably with a 47k resistor in series with the probe to limit the signal) and a high-resistance voltmeter to check supply, collector and emitter voltages, etc. The emitter current, and thus the collector current which is roughly the same, can be worked out from Ohm's law by dividing the voltage across R1 (Fig. 65) by the resistance of R1. Say the voltage across R1 is 0.5 V and R1 is 100 ohms, then the emitter current will be 0.5/100 A or 5 mA.

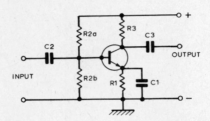

*Fig. 65. Typical transistor amplifier stage
to illustrate fault finding procedure*

A typical transistor amplifier stage is shown in Fig. 65. On checking, the collector current may be found to be zero, or lower or higher than it should be.

If it is zero, the emitter voltage will be zero and the collector voltage will equal the line voltage. If the base voltage is very low or zero, either R2a is open-circuit or R2b is short-circuit. If the base voltage is correct, R1 may be open-circuit. If R3 is open-circuit the emitter and collector voltages will be low.

If the collector current is low, the base-emitter voltage will also be low either because: (a) low base voltage because R2a is high or R2b low; (b) high emitter voltage because R1 is high or open-circuit.

If the collector current is high, the base-emitter voltage will be high either because: (a) high base voltage because R2a is low, R2b high or C2 leaky; (b) low emitter voltage because R1 is low or C1 leaky.

How can a silicon transistor be checked?

When a silicon transistor is conducting, the base voltage should be about 0.5 to 0.6 V higher than the emitter voltage. If the transistor is used as a linear amplifier, the collector voltage should be equal approximately to half of the supply voltage. A very useful rule-of-thumb concerning silicon transistors is that the collector current increases tenfold for each 60 mV (0.060 V) increase in base voltage. Another useful rule-of-thumb is that the gain of a linear amplifier is equal to 40 times the voltage across the collector load resistor (with no input signal).

What precautions should be taken when servicing transistorised equipment?

Never make resistance measurements or continuity tests with an ohmmeter giving an output voltage greater than 1.5 V; make sure that the soldering iron you use and all test gear are connected to a good common earth; switch off or disconnect the supply before removing any component; before replacing a transistor, check the associated components.

Switch off or disconnect the supply before soldering, and always use a heatsink with the soldering iron. Check transistor connections before connecting in circuit, and be sure not to short-circuit the transistor leads – this can easily be done when using a screwdriver or probe. It is most important when connecting the supply to observe correct polarity. Transistor leads should not be bent nearer than 1.5 mm from the seal. Remember that most transistors are light-sensitive and some are protected only by an opaque coating which must not be damaged. And remember that a faulty transistor may be very hot.

Do not use carbon tetrachloride or trichloroethylene alone for cleaning switching contacts or potentiometers as these cleaning agents tend to be corrosive.